BIOMIMETIC PROSTHETICS

Edited by **Ramana Vinjamuri**

Biomimetic Prosthetics
http://dx.doi.org/10.5772/66062
Edited by Ramana Vinjamuri

Contributors

Diana Alejandra Contreras Alejo, Francisco J. Gallegos-Funes, Raviraj Nataraj, Chiharu Ishii, Smita Nayak, Prasanna Lenka, Ebrahim A. Mattar, Hessa Al-Junaid, Hamad Al-Seddiqi, Ramana Vinjamuri

Published by InTech

Janeza Trdine 9, 51000 Rijeka, Croatia

Publishing Process Manager Edi Lipovic
Technical Editor SPi Global
Cover InTech Design team

First published February, 2018
Printed in Croatia
Legal deposit, Croatia: National and University Library in Zagreb

Additional hard copies can be obtained from orders@intechopen.com

Biomimetic Prosthetics, Edited by Ramana Vinjamuri
p. cm.
Print ISBN 978-953-51-3819-8
Online ISBN 978-953-51-3820-4

Meet the editor

Ramana Vinjamuri received his undergraduate degree in Electrical Engineering from Kakatiya University (India) in 2002. He received his MS degree in Electrical Engineering from Villanova University in 2004 specialized in bioinstrumentation. He received his PhD degree in Electrical Engineering in 2008 specialized in dimensionality reduction in control and coordination of human hand from the University of Pittsburgh. He worked as a postdoctoral fellow (2008–2012) in the field of brain-machine interfaces (BMI) to control prosthesis in the School of Medicine, University of Pittsburgh. He worked as a research assistant professor in the Department of Biomedical Engineering at the Johns Hopkins University (2012–2013). He is currently an assistant professor in Biomedical Engineering at the Stevens Institute of Technology. He also holds a secondary appointment as an adjunct assistant professor at the Indian Institute of Technology, Hyderabad, India. His research is supported by funding from private foundations and national institutes in the USA.

Contents

Preface

It was during my graduate studies, in 2005, that I was encouraged to work on a difficult problem of reducing the high dimensionality in control and coordination of the human hand. This is a complex problem as there are over 25 degrees of freedom in this complex biomechanical architecture of the human hand. How the brain controls the human hand was one of the difficult questions that fascinated me. In my graduate work, I developed theoretical and conceptual models about how intelligent higher level centers (brain and spinal cord) can control lower level biomechanical systems (limbs). Later, in 2008, in my postdoctoral work, I was introduced to brain-machine interfaces, where I got an opportunity to test my previously developed theoretical models in the real world and in real time. The applications were limitless in the areas of prosthetics, paralysis, and many more. During these years, I had the opportunity to work with individuals with spinal cord injury, stroke, and epilepsy.

In 2014, one life-changing event occurred in my life that shattered me. My mother, who was 61 at that time, had multiple seizures and stroke. Since then, she has right hand paralysis. The very hand that taught me how to write and how to draw pencil drawings is now dysfunctional. Her right hand is not functional to the extent that she cannot even perform basic activities of daily living on her own. She lost her independence. Her quality of life diminished in one fateful day. Although my area of research is in the field of assistive devices and prosthetics, I could not empathize more with the fact that there were very limited solutions that can bring back even half of her hand function without inconvenience. How can we, engineers and scientists, improve the feel and function of assistive devices? This question keeps me up at nights.

My research is inspired and encouraged by many individuals such as my mother who need assistive devices that can be more natural and less mechanistic. Biomimetic prosthetics is one such flourishing field of research that brings assistive technology to finger tips. This book is an effort to bring forth some important contributions in this field of research.

In this book, the readers will find various fields of biomimetic prosthetics in many areas of this research such as neural control of prosthetics, sensory feedback in prosthetics, user integration in assistive devices, aesthetics in prosthetics, and rehabilitation systems.

The introductory chapter briefly presents and puts biomimetic prosthetics in the context of current prosthetics. This chapter discusses different fields of research and provides brief update on ongoing research in upper and lower limb prosthetics and exoskeletons.

The second chapter talks about decoding high-dimensional hand movements using noninvasive electroencephalographic (EEG) signals. Decoding the motor intent from neural signals forms the basis of brain-machine interfaces. Understanding of neural basis of movement control enables us to design intuitive motor prosthesis that can mimic human behavior. Decoding high-dimensional hand movements was only possible in invasive brain recordings. Only recently researchers are attempting to decode hand movements in noninvasive brain recordings. In this chapter, by using fuzzy logic-based clustering and learning, hand movements were decoded from EEG signals.

In the thirdchapter, sensory feedback in prosthetics is discussed. While action from the brain to the end effector is the first half of the story, the rest belongs to sensation in the form of feedback from end effector fed back to the brain. Sensory feedback plays a vital role in the function of our limbs. In this chapter, a sensory feedback device for myoelectric prosthetic hand was reported that can enhance the quality of life of myoelectric prosthetic hand users. Two types of sensory feedback-force sense feedback and temperature sense feedback were presented here.

In the fourth chapter, an important aspect of user integration with assistive devices is presented. User integration with assistive devices to improve function is a key principle to consider to optimize the performance of prosthetics. This integration includes adapting and customizing operation of devices and training programs according to several user characteristics. These characteristics may be physical dimensions, residual capabilities, restored sensory feedback, cognitive perception, or stereotypical actions. Adapting the design to include these characteristics enables the user integration.

In the fifth chapter, the cosmetic prostheses that are natural in appearance are discussed. In addition to the function of prosthetic, form, and appearance, in other words, aesthetics are an important feature in improving the acceptability and usage of prosthetics. This chapter, although limited to passive prosthetics, presents an important contribution that could be used in conjunction with developments in active prosthetics.

In the sixth chapter, computer vision systems for tracking movements by prosthetic users are presented. The treatment of postdebilitating injuries is a continuous dynamic process that begins with prosthetic training and rehabilitation. Rehabilitation treatment can be carried out with computer vision systems such as recognition systems. This chapter discusses the development of a skin color recognition system to be used with the residual limb and prosthesis of skin-colored frame.

Finally, I would like to thank all the authors for their scholarly contributions and also IntechOpen staff especially Edi Lipovic, Mirena Calmic, and Ana Pantar for their kind assistance throughout the editing process. Without their help, this book would not be possible. I hope that readers will benefit in many ways from reading this book.

Ramana Vinjamuri
Assistant Professor
Department of Biomedical Engineering
Schaefer School of Engineering
Stevens Institute of Technology
New Jersey, USA

Chapter 1

Introductory Chapter: Toward Near-Natural Assistive Devices

Martin Burns, Michelle Schumacher and
Ramana Vinjamuri

Additional information is available at the end of the chapter

http://dx.doi.org/10.5772/intechopen.73637

1. Introduction

For thousands of years, the state-of-the-art prosthetics were very primitive. This is not due to lack of effort or expertise. Rather, the technological capabilities of these eras restricted the possible advancements in prosthetics, such that the most effective solutions leaned heavily toward the most basic designs. These historic prosthetics were devices made of wood, metal, and leather, crafted by blacksmiths and tradesmen, which served more as esthetic pieces than functional tools. Technology today has progressed to the point where biomimetic and anthropomorphic prosthetics are a reality and are beginning to rival their organic counterparts in form and function. Biomimetics, as coined by Otto Schmitt, is the concept in which an artificial device effectively mimics the structural, functional, and biological properties of the natural entity in which it is modeled after [39]. Alternatively, devices that are anthropomorphic in nature mimic the physical characteristics of the limb including its look and feel, integrating aspects such as a textured skin-like material onto the device. These concepts in prosthetic design are enabled by control systems, actuator designs, sensor and biosignal innovations, biomechanical insights, and battery technology, among others. These advances have also led to wearable robotic exoskeletons, allowing functional assistance for individuals with paralysis whose affected limb remains intact. Properly leveraging these technologies requires one to examine the device qualities desired by the user and learn about recent progress in research and development. In this book, we will discuss these issues in the context of assistive limb prosthetics and recent advances in the field.

2. Background

An estimated 185,000 people undergo an amputation every year in America [32]. In the United States, 1.6 million people were living with the loss of a limb in 2005, a number which is expected to more than double by the year 2050 [42]. There are numerous causes for limb loss, which can range from congenital-related instances to trauma [1]. The most common causes are dysvascular and trauma related, making up 82 and 16% of all amputations, respectively [18]. Of all trauma-related amputations, upper limb amputations specifically account for 69%, as well as 58% of all congenital-related amputations [3]. The economic cost to amputees comes primarily from the procedure itself, additional hospital bills, prosthetic devices, and lost wages. A traumatic lower limb amputation, as tracked in the NIH LEAP study, is expected to cost the amputee $350,465 over 40 years, with $91,105 incurred within the first 2 years of the injury. Of this total, an individual will spend $181,500 on prosthetics at approximately $10,232 per unit [13]. These costs are also corroborated in several unrelated studies on civilian and veteran lower limb amputation costs [5, 19]. Similar estimates for upper limb amputations are not readily available in the literature. Considering that children, along with veterans, are one of the leading recipients of prosthetics [3], an additional economic burden comes with the necessity of device abandonment. As most prosthetics are unable to change with a growing child, a new prosthetic must be purchased to match the size of the child [3, 43]. Device rejection rates in adolescents averaged approximately 45 and 35% for body-powered and electric prosthetics, while adult amputees saw significantly lower rejection rates of 26 and 23% for body-powered and electric prosthetics, respectively [5].

Spinal cord injury, traumatic brain injury, and stroke are other major causes of functional impairment. Traumatic brain injury alone was estimated to occur 1.74 million times per year in the United States, with 10% of those occurrences being classified as severe with long-term disability. Of those who are hospitalized for acute traumatic brain injury, 43% develop long-term disability and have difficulty completing activities of daily living (ADL). Instances of spinal cord injury occur between 12,000 and 20,000 times per year; 238,000–332,000 survivors are still alive in the United States today, all with varying degrees of impairment. Stroke is the most prevalent of the aforementioned conditions, with 795,000 annual occurrences, 610,000 of which are first-time strokes. One study found 26% of individuals with stroke required assistance for ADL, while half had reduced motor function and 30% required assistance walking. An individual who survives a stroke will see a lifetime cost of $140,000, while an individual with spinal cord injury will see anywhere from $334,170 to $1.02 million in expenses in the first year and $40,589 to $177,808 per year after. The outcomes of traumatic brain injury vary too much to estimate the annual cost per individual, but the overall direct cost in the United States has been estimated to be $9.2 billion per year and $51.2 billion in indirect costs of work and lost productivity [28].

An individual must undergo rehabilitation in the aftermath of any amputation or paralysis in order to maintain as much function as possible. Many researchers believe rehabilitation outcomes can be improved by using virtual reality-based training. This can be made more effective by tailoring difficulty of training to the user and by encouraging the user to transfer or empathize with the virtual model. Research groups have only recently begun investigating virtual reality as a viable means of improving rehabilitation outcomes. Small-scale pilot studies have shown that individuals with stroke improved functional scores in the Wolf Motor Function Test and Fugl-Meyer Assessment after playing commercial virtual reality video games daily for 10 consecutive days [30]. These findings have been corroborated by numerous small studies. No major study has taken place to examine this effect on a large patient population, however. An individualized virtual reality rehabilitation scheme has been implemented in a case study with a single hemiparetic individual poststroke with positive results [20]. Another group conducted individualized virtual rehabilitation for 21 individuals with stroke, although they did not report standard functional evaluation scores for their subjects [9]. In theory, a subject who empathizes with the virtual model or exhibits transference onto the virtual world could experience greater functional recovery than if they were training on an abstract game. This is not a novel concept: the rubber hand experiment successfully mapped sensations on a rubber hand to the user's visually occluded hand by matching the sensation to the opposite limb [6]. This technique has been successfully used to reduce phantom limb pain in the form of mirror therapy [11] and virtual visual feedback [29].

Research conducted within the past 30 years has provided insight into the most sought-after qualities of prosthetics. Designing systems, which satisfy these needs, are essential, as upper limb amputees typically use their prosthetic for more than 8 hours a day [24]. The needs of these users vary based on the type of amputation and type of prosthetic used, but several common factors are present among the different groups. Passive prosthetic users tend to focus more on cosmetics, comfort, and basic function, while users of body-powered and myoelectric prosthetics tend to focus more on advanced function. Myoelectric prosthetic users tend to request sensory feedback, along with simpler controls that can provide more dexterous motions. Active or powered prostheses focus on even more advanced motor function by providing external power through actuated motors. The most universally desired characteristics among all groups are reductions in cost and weight. Overall, ADL tasks that people rank as most important include mobility, object handling, and manipulation [14].

The decision to choose a passive or active prosthetic is primarily motivated by the type of disability and the user's desired level of physical activity. Additional factors include the user's ability to operate a powered device and personal requirements for cost, weight, and reliability. A passive leg prosthetic would partially restore the user's mobility, whereas a passive upper limb prosthetic would not restore the user's ability to interact with objects to the same degree. Therefore, a passive lower limb prosthetic would be more appealing from a functional perspective than an equivalent upper limb prosthetic. Someone who expects minimal or low levels of physical activity would be more likely to choose a passive prosthetic, since the extra functionality of active systems would go unused while incurring additional cost and weight. Furthermore, complex active prosthetics are more likely to break down and need repair, which may not be acceptable for a user who is primarily interested in cosmetics.

There are several options for passive prosthetics. Local prosthetists are skilled in crafting custom-fit cuffs and limb components, while several companies specialize in producing convincing passive limbs. Some of the more prominent companies in upper limbs include Touch

Bionics and Liberating Technologies Inc. (LTI), which provide the Livingskin and FHL line of passive hands, respectively. Livingskin products are custom-made silicon prosthetics that are made to match the user's skin tone and markings and can act as stand-alone prosthetics or skin coverings for robotic systems. The FHL line of hands by LTI is foam endoskeletal hands, which can be posed and mounted onto other products offered by the company, including passive wrist, elbow, and shoulder prosthetics. Ottobock offers a whole-arm endoskeletal prosthetic consisting of a metal articulated skeleton encased in foam and a silicon skin. Ottobock also provides multiple passive lower limb prosthetics such as the Harmony P4, a vacuum-cuffed below knee passive ankle, and the Aqualine, an above-knee endoskeletal prosthetic, which features metals and coatings resistant to water and chlorine corrosion. The Solid Ankle, Cushioned Heel (SACH) foot produced by NZALS is a popular passive foot prosthetic featuring rubber regions that flex and deform through the wearer's gait, replicating how the foot articulates during normal walking. Össur produces several passive lower limb solutions for knee and ankle prosthetics for both low-activity and athletic populations. The Flex-Run and Flex-Foot Cheetah foot prosthetics by Össur are famous for providing unprecedented athletic capability for transtibial amputees.

An alternative to passive limbs is body-powered prosthetics, which is controlled through a series of cables that link the movement of the body to the movement of the prosthetic. These devices are built to perform specific tasks and have terminal devices such as hooks to perform gripping. However, these devices often require the user to exert large cable operation forces. A study performed by Hichert et al. found that the force required to activate multiple body-powered prosthetics elicited fatigue, discomfort, and even pain in the user [22]. One way to control the prosthetic without fatiguing the user is to use an active prosthetic, which is powered by external power sources and is capable of more precise movements through the use of actuators. Many active prosthetics adapt biosignal-based control, in which signals generated by the user's own body are utilized to control the actuation of the device. Translating these initial signals into precise control of the prosthetic device can vary depending on the individual needs of the particular prosthetic but generally includes three major stages: signal acquisition, signal processing, and prosthetic interfacing and actuation. Signal acquisition is dependent on the type of biosignal being measured and is subsequently converted from an analog to digital signal. From here, the signal can undergo processing, which consists of feature extraction (filtering and isolation of the targeted signal) and translation (determining the intended movement). The prosthetic interface then takes this information and signals the prosthetic to physically carry out the intended movement.

There are a few different biosignal inputs that can be used in active prosthetic control. Many current active prosthetics are myoelectric, which are controlled through the use of electromyography (EMG) signals, allowing the device to better reflect the intention of the user [21]. EMG signals have been utilized to successfully control prosthetic hands [37]. Some myoelectric devices have even been noted as commercially available and reimbursed by the French health-care system [27] The I-Limb Quantum from Touch Bionics by Össur is a current commercial model of a prosthetic hand that features 36 different available grip postures and can be controlled through three methods in addition to muscle-based control. The LUKE arm by Mobius Bionics is the first bionic arm to received market approval by the FDA. This arm is controlled through a variety of methods including EMG electrodes and features 10 degrees of freedom. A few drawbacks to muscle-based control technology, however, are its generally high cost [7] as well as repeatability and reliability issues in surface EMG signals [36], making this input method difficult to use in prosthetic control. Implanted EMG sensors are a promising technology that can provide high-quality signals; however, they are costly, invasive, and only in the primary stages of human use [33]. Myoelectric control also relies on undamaged neuromuscular pathways being accessible to present the EMG signal [7]. Although strides are being taken in targeted muscle reinnervation (TMR) to allow for greater EMG accessibility [21], this may not be possible in all cases. Electroencephalogram (EEG)-based control of prosthetics attempts to bypass this problem [21]. EEG-controlled prosthetics employ the user's brain signals in conjunction with brain-computer interface (BCI) systems for control. Brain signals have been successfully detected through the use of an EEG to effectively control prosthetics including that of bionic arms. Current bionic hands have been able to demonstrate two degrees of freedom in the fingers [4, 7]. Additionally, Vidaurre et al. demonstrated EEG-based control of an exoskeleton [38]. Invasive brain-computer interfaces have demonstrated accurate and real-time control of advanced prosthetics such as modular prosthetic limb [15, 23, 40, 41].

Restoring hand function is one of the top priorities for individuals with paralysis and other motor disabilities. Normal hand movement is dependent on both visual and tactile feedbacks to assist the brain in determining the proper hand grasp, grip strength, movement speed, etc. required to successfully perform the intended task. This feedback creates a closed-loop system between the limb and the brain. Most current prostheses, however, operate through open-loop control by which the flow of information is unidirectional, from the user to the prosthetic. It is believed that the implementation of feedback to create a closed-loop system will improve the performance of the prosthetic. In active prosthetics, three of the major ways to provide sensor feedback to the user include substitution, modality-matched, and somatically matched feedback mechanisms [35]. In general, substitution feedback is the application of an indirect stimulus that is intended to substitute another form of stimulus, i.e., substituting a touch stimulus with an auditory one. There are different types of feedback including vibrotactile and electrotactile sensory substitution, which utilize vibration and electrical current stimuli, respectively. Alternatively, modality-matched feedback applies an identical sensation to the one being detected but on a different part of the body. For instance, the user may feel a prosthetic hand's grip force through a force generated on the abdomen or chest. Somatotopically matched feedback causes the user to perceive the correct sensation from the correct location, as if the prosthetic was their natural limb. In order to create a prosthetic that most accurately performs sensory feedback, both modality-matched and somatotopically matched feedbacks should be employed [2, 35].

Somatotopically matched feedback may be made possible as an added benefit of afferent TMR [36]; however, damage to the peripheral nerves from amputation may have altered their ability to elicit the correct sensory response in thus altering the effects of the feedback [17]. In light of this, new strides are being taken to directly stimulate the human somatosensory cortex. A study conducted by Cronin et al. [16] demonstrated a subject's ability to modulate hand motor behavior in response to cortical sensory stimulation delivered by ECoG electrodes. It is believed that this same technology may be used in the tactile feedback of a prosthetic hand

[16]. Another very important factor to take into account, in addition to the type of feedback present, is the time delay of the stimulus. At a distance of approximately 1 meter, the conduction times of afferent nerves range between approximately 14 and 28 ms [25]; therefore a short latency of the additional feedback is important to create real-time reactions. Short latencies, similar to that of a normal human hand, are also believed to help promote user ownership of the prosthetic [35].

Prosthetic devices are traditionally viewed as systems intended to replace a missing limb. However, a prosthetic can also be seen as a device that restores a missing bodily function. Taking this definition introduces assistive robots such as exoskeletons, which promise to restore missing function in those who are partially or completely paralyzed in one or more parts of their body. Several companies offer exoskeletons for assistance and rehabilitation of individuals with stroke, including Myomo's MyoPro hand/arm system, Bioservo's SEM Glove, Hocoma's ArmeoPower and ManovoPower upper limb systems, and Rehab-Robotics' Hand of Hope. Cyberdyne's HAL Lower Limb, Rewalk's Rewalk Personal and Rewalk Rehabilitation, and Rex Bionics' Rex P among many others provide lower limb rehabilitation and walking platforms for those with paralysis. Exoskeletons also exist in the academic research field, including ExoGlove Poly [26], the Wyss Institute's Soft Robotic Glove [34], HandSOME [12], Pisa/IIT SoftHand [10], and several others [8, 31], which serve as platforms to research novel control algorithms.

The above discussion provides a brief background and establishes relevance to the topics discussed in this book. This discussion is in no way completely exhaustive. There are numerous research studies in many of these research areas that could not be discussed due to time and space constraints. In fact, there are many manuscripts in preparation and press while we are reading or writing this chapter. We believe that biomimetic prosthetics will play a major role in revolutionizing the form and function of assistive devices in the near future.

Author details

Martin Burns, Michelle Schumacher and Ramana Vinjamuri*

*Address all correspondence to: ramana.vinjamuri@gmail.com

Department of Biomedical Engineering, Stevens Institute of Technology, NJ, USA

References

[1] Limb Loss Task Force/Amputee Coalition. Roadmap for Preventing Limb Loss in America: Recommendations From the 2012 Limb Loss Task Force. Knoxville, Tennessee; 2012

[2] Antfolk C, D'Alonzo M, Rosén B, Lundborg G, Sebelius F, Cipriani C. Sensory feedback in upper limb prosthetics. Expert Review of Medical Devices. 2013;**10**(1):45-54. DOI: https://doi.org/10.1586/erd.12.68

[3] Baker C, Brent D, Wilson C, Xu J, Thompson LA. Additive Manufacturing for Economical, User-Accessible Upper-Limb Prosthetics. Prosthetics and Orthotics Open Journal. 2017;**1** (1,8):1-5 Available from: http://www.imedpub.com/articles/additive-manufacturing-for-economical-useraccessible-upperlimb-prosthetics.pdf

[4] Beyrouthy T, Al Kork SK, Akl Korbane J, Abdulmonem A. EEG mind controlled smart prosthetic arm. In: 2016 IEEE International Conference on Emerging Technologies and Innovative Business Practices for the Transformation of Societies. Mauritius: EmergiTech; 2016. pp. 404-409. DOI: https://doi.org/10.1109/EmergiTech.2016.7737375

[5] Biddiss E, Chau T. Upper limb prosthesis use and abandonment: A survey of the last 25 years. Prosthetics and Orthotics International. 2007;**31**(3):236-257. DOI: https://doi.org/10.1080/03093640600994581

[6] Botvinick M, Cohen J. Rubber hands 'feel' touch that eyes see. Nature. 1998;**391**(6669):756. DOI: https://doi.org/10.1038/35784

[7] Bright D, Nair A, Salvekar D, Bhisikar S. EEG-based brain controlled prosthetic arm. In: Conference on Advances in Signal Processing. Pune, India: CASP 2016; 2016. pp. 479-483. DOI: https://doi.org/10.1109/CASP.2016.7746219

[8] Burns MK, Van Orden K, Patel V, Vinjamuri R. Towards a wearable hand exoskeleton with embedded synergies. In: Proceedings of the Annual International Conference of the IEEE Engineering in Medicine and Biology Society. Jeju Island, Korea: EMBS; 2017. pp. 213-216. DOI: https://doi.org/10.1109/EMBC.2017.8036800

[9] Cameirao MS, Badia SB i, Oller ED, Verschure PFMJ. Neurorehabilitation using the virtual reality based rehabilitation gaming system: Methodology, design, psychometrics, usability and validation. Journal of Neuroengineering and Rehabilitation. 2010;**7**(1):48. DOI: https://doi.org/10.1186/1743-0003-7-48

[10] Catalano MG, Grioli G, Farnioli E, Serio A, Piazza C, Bicchi A. Adaptive synergies for the design and control of the Pisa/IIT SoftHand. The International Journal of Robotics Research. 2014;**33**(5):768-782

[11] Chan BL, Witt R, Charrow AP, Magee A, Howard R, Pasquina PF, Heilman KM, Tsao JW. Mirror therapy for phantom limb pain. New England Journal of Medicine. 2007;**357**(21): 2206-2207. DOI: https://doi.org/10.1056/NEJMc071927

[12] Chen J, Nichols D, Brokaw EB, Lum PS. Home-based therapy after stroke using the hand spring operated movement enhancer (HandSOME). IEEE Transactions on Neural Systems and Rehabilitation Engineering. 2017;**25**(12):1-1. DOI: https://doi.org/10.1109/TNSRE.2017.2695379

[13] Chung KC, Saddawi-Konefka D, Haase SC, Kaul G. A cost-utility analysis of amputation versus salvage for Gustilo type IIIB and IIIC open Tibial fractures. Plastic and Reconstructive Surgery. 2009;**124**(6):1965-1973. DOI: https://doi.org/10.1097/PRS.0b013e3181bcf156

[14] Cordella F, Ciancio AL, Sacchetti R, Davalli A, Cutti AG, Guglielmelli E, Zollo L. Literature review on needs of upper limb prosthesis users. Frontiers in Neuroscience. 2016;**10** (May):1-14. DOI: https://doi.org/10.3389/fnins.2016.00209

[15] Collinger JL, Wodlinger B, Downey JE, Wang W, Tyler-Kabara EC, Weber DJ, McMorland AJC, Velliste M, Boninger ML, Schwartz AB. High-performance neuroprosthetic control by an individual with tetraplegia. The Lancet. 2013;**381**(9866):557-564

[16] Cronin JA, Jing W, Collins KL, Sarma D, Rao RPN, Ojemann JG, Olson JD. Task-specific somatosensory feedback via cortical stimulation in humans. IEEE Transactions on Haptics. 2016;**9**(4):515-522. DOI: https://doi.org/10.1109/TOH.2016.2591952

[17] D'Anna E, Petrini FM, Artoni F, Popovic I, Simanić I, Raspopovic S, Micera S. A Somatotopic bidirectional hand prosthesis with transcutaneous electrical nerve stimulation based sensory feedback. Scientific Reports. 2017;**7**(1). DOI: https://doi.org/10.1038/s41598-017-11306-w

[18] Dillingham TR, Pezzin LE, Mackenzie EJ. Limb amputation and limb deficiency: Epidemiology and recent trends in the United States. Southern Medical Journal. 2002;**95**(8):875-884. DOI: https://doi.org/10.1097/00007611-200208000-00018

[19] Edwards DS, Phillip RD, Bosanquet N, Bull AMJ, Clasper JC. What is the magnitude and long-term economic cost of care of the British Military Afghanistan Amputee Cohort? Clinical Orthopaedics and Related Research. Springer US. 2015;**473**(9):2848-2855. DOI: https://doi.org/10.1007/s11999-015-4250-9

[20] Fluet GG, Merians AS, Qiu Q, Saleh S, Ruano V, Delmonico AR, Adamovich SV. Robotic/virtual reality intervention program individualized to meet the specific sensorimotor impairments of an individual patient: A case study. International Journal on Disability and Human Development. 2014;**13**(3):401-407. DOI: https://doi.org/10.1515/ijdhd-2014-0334

[21] Geethanjali P. Myoelectric control of prosthetic hands: State-of-the-art review. Medical Devices: Evidence and Research. 2016;**9**:247-255. DOI: https://doi.org/10.2147/MDER.S91102

[22] Hichert M, Vardy AN, Plettenburg D. Fatigue-free operation of most body-powered prostheses not feasible for majority of users with trans-radial deficiency. Prosthetics and Orthotics International. 2017;**0**(0):309364617708651. DOI: https://doi.org/10.1177/0309364617708651

[23] Hotson G, McMullen DP, Fifer MS, Johannes MS, Katyal KD, Para MP, Armiger R, et al. Individual finger control of a modular prosthetic limb using high-density electrocorticography in a human subject. Journal of Neural Engineering. 2016;**13**(2):026017

[24] Jang CH, Yang HS, Yang HE, Lee SY, Kwon JW, Yun BD, Choi JY, Kim SN, Jeong HW. A survey on activities of daily living and occupations of upper extremity amputees. Annals of Rehabilitation Medicine. 2011;**35**:907. DOI: https://doi.org/10.5535/arm.2011.35.6.907

[25] Johansson RS, Flanagan JR. Coding and use of tactile signals from the fingertips in object manipulation tasks. Nature Reviews Neuroscience. 2009;**10**(5):345-359. DOI: https://doi.org/10.1038/nrn2621

[26] Kang BB, Lee H, In H, Jeong U, Chung J, Cho KJ. Development of a polymer-based tendon-driven wearable robotic hand. In: Proceedings – IEEE International Conference on Robotics and Automation, June 2016. 2016. pp. 3750-3755. DOI: https://doi.org/10.1109/ICRA.2016.7487562

[27] Loiret I, Sanamane V, Touillet A, Martinet N, Paysant J, Fournier-Farley C, François A-G. Assessment of multigrip prosthetic hand by a crossover longitudinal study. Annals of Physical and Rehabilitation Medicine. 2017;**60**(Suppl):e34. DOI: https://doi.org/https://doi.org/10.1016/j.rehab.2017.07.005

[28] Ma VY, Chan L, Carruthers KJ. The incidence, prevalence, costs and impact on disability of common conditions requiring rehabilitation in the US: Stroke, spinal cord injury, traumatic brain injury, multiple sclerosis, osteoarthritis, rheumatoid arthritis, limb loss, and back pain. Archives of Physical Medicine and Rehabilitation. 2014;**95**(5):986-995. DOI: https://doi.org/10.1016/j.apmr.2013.10.032

[29] Mercier C, Sirigu A. Training with virtual visual feedback to alleviate phantom limb pain. Neurorehabilitation and Neural Repair. 2009;**23**(6):587-594. DOI: https://doi.org/10.1177/1545968308328717

[30] Mouawad MR, Doust CG, Max MD, McNulty PA. Wii-based movement therapy to promote improved upper extremity function post-stroke: A pilot study. Journal of Rehabilitation Medicine. 2011;**43**(6):527-533. DOI: https://doi.org/10.2340/16501977-0816

[31] Nycz CJ, Butzer T, Lambercy O, Arata J, Fischer GS, Gassert R. Design and characterization of a lightweight and fully portable remote actuation system for use with a hand exoskeleton. IEEE Robotics and Automation Letters. 2016;**1**(2):976-983. DOI: https://doi.org/10.1109/LRA.2016.2528296

[32] Owings MF, Kozak LJ, Ambulatory and inpatient procedures in the United States, 1996. In: Vital and Health Statistics, Series 13, Data from the National Health Survey, No. 139. 1998. pp. 1-119

[33] Pasquina PF, Evangelista M, Carvalho AJ, Lockhart J, Griffin S, Nanos G, McKay P, et al. First-in-man demonstration of a fully implanted myoelectric sensors system to control an advanced electromechanical prosthetic hand. Journal of Neuroscience Methods. 2015;**244**: 85-93. DOI: https://doi.org/10.1016/j.jneumeth.2014.07.016

[34] Polygerinos P, Wang Z, Galloway KC, Wood RJ, Walsh CJ. Soft robotic glove for combined assistance and at-home rehabilitation. Robotics and Autonomous Systems. Elsevier BV. 2015;**73**:135-143. DOI: https://doi.org/10.1016/j.robot.2014.08.014

[35] Schofield JS, Evans KR, Carey JP, Hebert JS. Applications of sensory feedback in motorized upper extremity prosthesis: A review. Expert Review of Medical Devices. 2014;**11**(5): 499-511. DOI: https://doi.org/10.1586/17434440.2014.929496

[36] Schultz AE, Kuiken TA. Neural interfaces for control of upper limb prostheses: The state of the art and future possibilities. PM & R: The Journal of Injury, Function, and Rehabilitation. 2011;**3**(1):55-67. DOI: https://doi.org/10.1016/j.pmrj.2010.06.016

[37] Tavakoli M, Benussi C, Lourenco JL. Single channel surface EMG control of advanced prosthetic hands: A simple, low cost and efficient approach. Expert Systems with Applications. 2017;**79**:322-332. DOI: https://doi.org/10.1016/j.eswa.2017.03.012

[38] Vidaurre C, Klauer C, Schauer T, Ramos-Murguialday A, Müller KR. EEG-based BCI for the linear control of an upper-limb neuroprosthesis. Medical Engineering and Physics. 2016;**38**(11):1195-1204. DOI: https://doi.org/10.1016/j.medengphy.2016.06.010

[39] Vincent JFV, Bogatyreva OA, Bogatyrev NR, Bowyer A, Pahl A-K. Biomimetics: Its practice and theory. Journal of the Royal Society Interface. 2006;**3**(9):471-482. DOI: https://doi.org/10.1098/rsif.2006.0127

[40] Vinjamuri R, Weber DJ, Mao Z-H, Collinger JL, Degenhart AD, Kelly JW, Boninger ML, Tyler-Kabara EC, Wang W. Toward synergy-based brain-machine interfaces. IEEE Transactions on Information Technology in Biomedicine. 2011;**15**(5):726-736

[41] Wang W, Collinger JL, Degenhart AD, Tyler-Kabara EC, Schwartz AB, Moran DW, Weber DJ, et al. An electrocorticographic brain interface in an individual with tetraplegia. PLoS One. 2013;**8**(2):e55344

[42] Ziegler-Graham K, MacKenzie EJ, Ephraim PL, Travison TG, Brookmeyer R. Estimating the prevalence of limb loss in the United States: 2005 to 2050. Archives of Physical Medicine and Rehabilitation. 2008;**89**(3):422-429. DOI: https://doi.org/10.1016/j.apmr.2007.11.005

[43] Zuniga J, Katsavelis D, Peck J, Stollberg J, Petrykowski M, Carson A, Fernandez C. Cyborg beast: A low-cost 3D-printed prosthetic hand for children with upper-limb differences. BMC Research Notes. 2015;**8**(1):10. DOI: https://doi.org/10.1186/s13104-015-0971-9

Chapter 2

Biomimetic Based EEG Learning for Robotics Complex Grasping and Dexterous Manipulation

Ebrahim A. Mattar, Hessa J. Al-Junaid and Hamad H. Al-Seddiqi

Additional information is available at the end of the chapter

http://dx.doi.org/10.5772/intechopen.72455

Abstract

There have been tremendous efforts to understand the biological nature of human grasping, in such a way that it can be learned and copied to prosthesis–robotics and dextrous grasping applications. Several biomimetic methods and techniques have been adopted, hence applied to analytically comprehend ways human performs grasping to duplicate human knowledge. A major topic for further study, is related to decoding the resulting EEG brainwaves during motorizing of fingers and moving parts. To accomplish this, there are a number of phases that are performed, including recording, pre-processing, filtration, and understanding of the waves. However, there are two important phases that have received substantial research attentions. The classification and decoding, of such massive and complex brain waves, as they are two important steps towards understanding patterns during grasping. In this respect, the fundamental objective of this research is to demonstrate how to employ advanced pattern recognition methods, like fuzzy c-mean clustering for understanding resulting EEG brain waves, in such a way to control a prosthesis or robotic hand, while relying sets of detected EEG brainwaves. There are a number of decoding and classification methods and techniques, however we shall look into fuzzy based clustering blended with principle component analysis (PAC) technique to help for the decoding mechanism. EEG brainwaves during a grasping and manipulation have been used for this analysis. This involves, movement of almost five fingers during a grasping defined task. The study has found that, it is not a straight forward task to decode all human fingers motions, as due to the complexity of grasping tasks. However, the adopted analysis was able to classify and identify the different narrowly performed and related fundamental events during a simple grasping task.

Keywords: biomimetic, electroencephalography, PCA, fuzzy clustering, Dextrous grasping, prosthesis, robotic hand, robotics control

1. Introduction

1.1. Biomimetic engineering for EEG applications

Human brain and other biological brains are the most complicated part of any intelligent life forms, it controls almost every part in the human body. For controlling biological organs via a brain, there must be communication between organs. After years of research, it was established that brains communicate with organs via massive set of electrical signals that are periodically sent to various body parts. These electrical signals, can be detected through electroencephalography (EEG). These EEG waves are sent from neurons in the brain to the ones in the spinal cord, and end in nerve endings in the organs. Neurons are connected together, in addition to nerves using synapses. This is the basic structure of the nervous system that is in charge of communications between the brain and the rest of the body. Brain signals are basically electrical currents, however connecting them to artificial prosthesis, and robotic hand has always been a challenge.

Given this fact, biomimetics is employed for imitation of models, systems, in addition to elements of nature, Vincent et al. [1]. This is used for a purpose of solving complex human related problems. Lately, electroencephalography(EEG)–has received considerable attention, due to a number of advantages while data mining and dealing with brain waves. There are tremendous efforts to make use of biomimicry for EEG waves understanding, [2–5]. In this regard, Yuanfang et al. [6], stated they have used two novel classifiers: biomimetic pattern recognition and sparse representation. Each classifier labels the unlabelled data for other classifiers to extend the labeled set. An enlarged labeled set is used to generate the final classifier BPR–SR. BPR–SR was constructed by combining BPR with SR. EEG features extracted by common spatial patterns were used for classification. Perruchoud et al. [7], presented the concept of biomimetic rehabilitation engineering, and presented a more focus review in this sense. The importance of somatosensory feedback for brain–machine interfaces were also presented. Menniti et al. [8] stated that they used implanted electrodes, to obtain acoustical information about the external environment generated by a biomimetic system and converted in electrical signals. Bullock et al. [9], data was collected by recording two male experienced machinist and two female housekeepers while working for at least 8 hours each. Classification was done using modified Feix taxonomy to work with video footage. Data showed that 80% of the time the housekeepers used only five grasps and the machinists 10 grasps. In [10] Bullock et al., it was found out that for basic objects handling, medium wrap and lateral pinch are most suited. For precision and dexterous manipulation, three fingertips were needed and thumb −two finger, tripod, or lateral tripod can be used.

In [11], two machinists and two housekeepers were recorded during typical day at work. A camera was head-mounted and was used to record hands when doing their work. With each right hand grasp, the data was tagged with grasp type and object properties. In Feix et al. [12], analysis of human grasping and behavior detection was done. Object characteristics and grasp types were reported. Mainly, 10,000 grasp instances were recorded and a correlation between the property of the object and the grasp type was established. In reference to [13, 14], various analysis of EEG were conducted using some time and frequency analysis approaches. In [15],

http://dx.doi.org/10.5772/intechopen.72455

Feix et al., investigated task classification according to degrees of freedom, required force, and the functional task type. Since these are thought to be natural considerations that humans think of when doing tasks. In [16–18], have presented various ways for EEG analysis. A versatile signal processing and analysis framework for EEG were conducted. Signals were decomposed into frequency sub-bands using *DWT*. Set of statistical features were extracted from the sub-bands to represent distribution of wavelet coefficients. Hazrati and Erfanian [19], have presented an online EEG-based brain–computer interface for controlling hand grasp using an adaptive probabilistic ANN. Jianjun et al. [20], have also implemented a robotics control through a set of EEG waves. Group of 13 human subjects willingly modulate brain activity to control a robotic arm.

Furthermore, due to complex robotic tasks, grasping by dextrous prosthesis–robotic hands, is no longer can be achieved through analytical approaches, as done in the past using mathematically defined force computations, as found in Vinet et al. [21], and Cutkosky [22]. However, human grasping experience and knowledge have been under massive studies to be transferred to robots, Mattar et al. [23]. This is due to the advanced applications robots and prosthesis are preforming nowadays.

1.2. Electroencephalography classification-decoding

In addition to the above introduction, recently human like hands have been developed world-wide, to achieve accurate grasping, **Figure 1**, [24, 25]. In parallel, EEG based robotics control systems have also received fundamental attention by researchers world-wide, Vinet et al. [21]. However, there are a number of issues related to interpretation of the massive waves. EEG signal classification using principal component analysis (PCA), independent component analysis (ICA), linear discriminant analysis (LDA) and SVM, were investigated by Subasi and Gursoy [26]. In their work, they have investigated using EEG in diagnostics of epileptic seizure. First, data was acquired from an outside source. These data contained EEG recordings from multiple seizure patients that have been diagnosed and found the region where the seizure accrued and another data from healthy people. The EEG signals then were transformed to sub-bands using discrete wavelet transform (DWT). DWT was used because the EEG signals are non-stationary and DWT is capable of handling such signals. Hence using statistical techniques, features related to

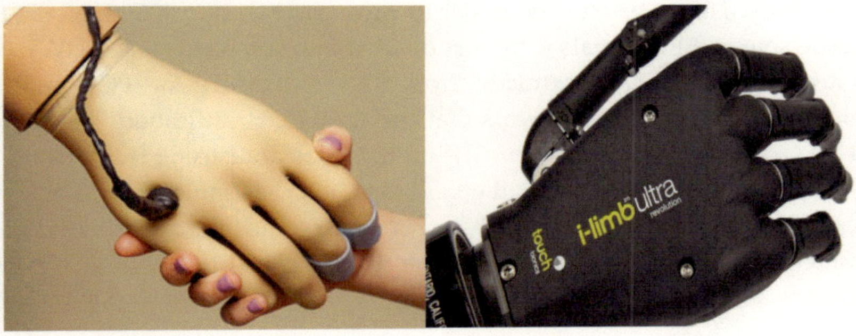

Figure 1. Current development of prosthesis, and robotic artificial limbs and hand. *Left picture source with permission,* Dustin, [24]. *Right source with permission,* Touch Bionics, [25].

seizures were extracted. These techniques reduce the data by removing the redundancy in it and focusing on the features. This is done by getting the low dimensional space of the features. The first technique used was principal component analysis (PCA) that reduces dimension and maximize variance in the data of the lower dimension hence make it easier to extract waves features. Independent component analysis (ICA) was used. This method generates mutually independent components from random signals. This highlights the features in the signals. Finally, linear discriminant analysis (LDA) was used. This method combines the predictors to generate a discriminant score. This resulted in discriminant scores normally distributed in each class. All the features from each technique was then submitted to support vector machine (SVM). The SVM then generated the classification of having an epileptic seizure or not. The accuracy of feature extraction was based on the specificity and sensitivity derived from the confusion matrices. The PCA had the lowest classification at 98.75%. ICA achieved second best result at 99.5%. LDA achieved a perfect score of 100%. The research showed that the generalization performance of SVM was improved by dimensional reduction. This whole system was proven to be eligible for the use in diagnostics.

Wang et al. [27], have also shown how feature extraction and recognition of epileptiform activity in EEG by combining PCA with approximate entropy (ApEn) can be adopted. The paper also investigated the use of EEG signals to identify the epileptiform activity associated with epilepsy. The problems with older techniques for finding epileptiform activity was the speed, mainly due to the huge amounts of data. First, the EEG data was recorded and then was subjected to PCA to reduce the dimension of data, hence to identify patterns related to epileptiform activity based on the variances. DWT was applied on the PCA data to generate the sub-bands associated with signal types such as spike and sharp waves. Then, approximate entropy (ApEn) estimation was done of the sub-bands. The ApEn is used to find out if the time series is random or deterministic based on the value of the entropy (high or low respectively). Finally, to classify the data into epilepsy or not, Neyman-Pearson criteria was applied. This was done by getting a threshold value from the Neyman-Pearson criteria and comparing the ApEn value with the threshold. If the signal ApEn was less than a threshold, it is epileptic otherwise it is normal. The paper showed result is similar to other techniques, however because of PCA the detection speed was much faster due to reduction of data. Mahajan et al. [28] have done classification of EEG using PCA, ICA, and neural network. This is also another paper that focuses on the use of EEG signals to diagnose epilepsy. This research was similar to Subasi and Gursoy, [26]. EEG data was taken from an outside source. EEG signals were first divided into sub-bands using DWT. Then, some features related to epilepsy were extracted from the sub-bands using PCA and ICA. Then, features were sent to train a neural network that was used for classification. The neural network is made of two layers and five perceptron. The training was done using feedforward algorithm. The neural network was able to classify the EEG data and determine whether it is epileptic or not. The accuracy of feature extraction was based on the specificity and sensitivity. ICA achieved 96.75% accuracy and PCA achieved 93.63%. The work showed that ICA is better than PCA which was also established in by Subasi and Gursoy.

For electroencephalogram signal classification using neural networks with wavelet packet analysis, PCA and data normalization as pre-processors was introduced by Aminian et al. [29]. In this work, the authors try to predict hand/arm movement using EEG data. The EEG signals were recorded when subjects were asked to grasp an object, point to an object, and extend an arm. This data was then subjected to wavelet packet analysis to reduce the number of features to concentrate on the important ones. This was done by choosing approximate coefficients as features. Then, generates an approximation and detail levels from the each of the approximation and detail component by decomposing it. PCA is used to reduce features to reduce the amount of data while keeping the important parts. A uniform scaling scheme was applied to make sure that small important features are not taken for granted. This was achieved by normalization of the data by getting zero mean and unity standard deviation. Finally, a neural network was used for classification. It is a backpropagation trained neural network with feedforward multilayers. The neural network generates outputs as many as the classes (tasks). By doing so, each output corresponds to one of the classes and gives probabilities for the inputs and their class. The network was unaffected by outliers because it does not try to fit them due to its simplicity. Overtraining was also not an issue because of smooth mapping function of the input–output. This scheme achieved a 100% accuracy in the classification of the tasks for all the participants. Classification of direction perception EEG based on *PCA-SVM* was also introduced by Wang and Wang, [30]. In this work, the paper uses EEG signals to identify movement based on vision. The *EEG* data was recorded when subjects were looking at a monitor with 3D environment. This environment contained roads and there was movement in first person view. The objective was to identify the direction of movement perceived by these individuals.

Discrete Fourier transform (DFT) was first applied to these signals to examine them in the frequency domain. This allowed the identification of some features related to the perception of movement based on the power at certain frequencies. Then, PCA was used to reduce the dimension of the data and simplify the features. To classify the data, support victor machine SVM was used. To overcome some error induced by the penalty factor in SVM, a new variable was introduced to add degree of freedom margin in patter recognition. The two class classification achieved (50–91.25%) accuracy between all the subjects and the two way movement (*left-right, right-front…*). This could be due to the feature extraction technique used which is not able to distinguish between similar movements. The use of DWT may have been a better choice due its ability to handle non-stationary signals such as EEG.

Within this chapter, and given the above review, the adopted technique, we shall introduce another view for EEG analysis for prosthesis–robotics grasping. This is based on blending PCA with C-means clustering. The technique has an ability to detect the grasping events.

1.3. Chapter contribution and organization

1.3.1. Chapter contribution

This chapter is presenting a unique classification technique for EEG brainwaves during a defined grasping and lifting task. The study is presenting a mechanism for extracting focal features, and patterns intrinsics for resulting EEG waves during experimentally conducted grasping task. Definitely, knowledge of the features will help for robotics grasping based tasks. The novel presented work in this chapter, is summarized in **Figure 2**, and as follow:

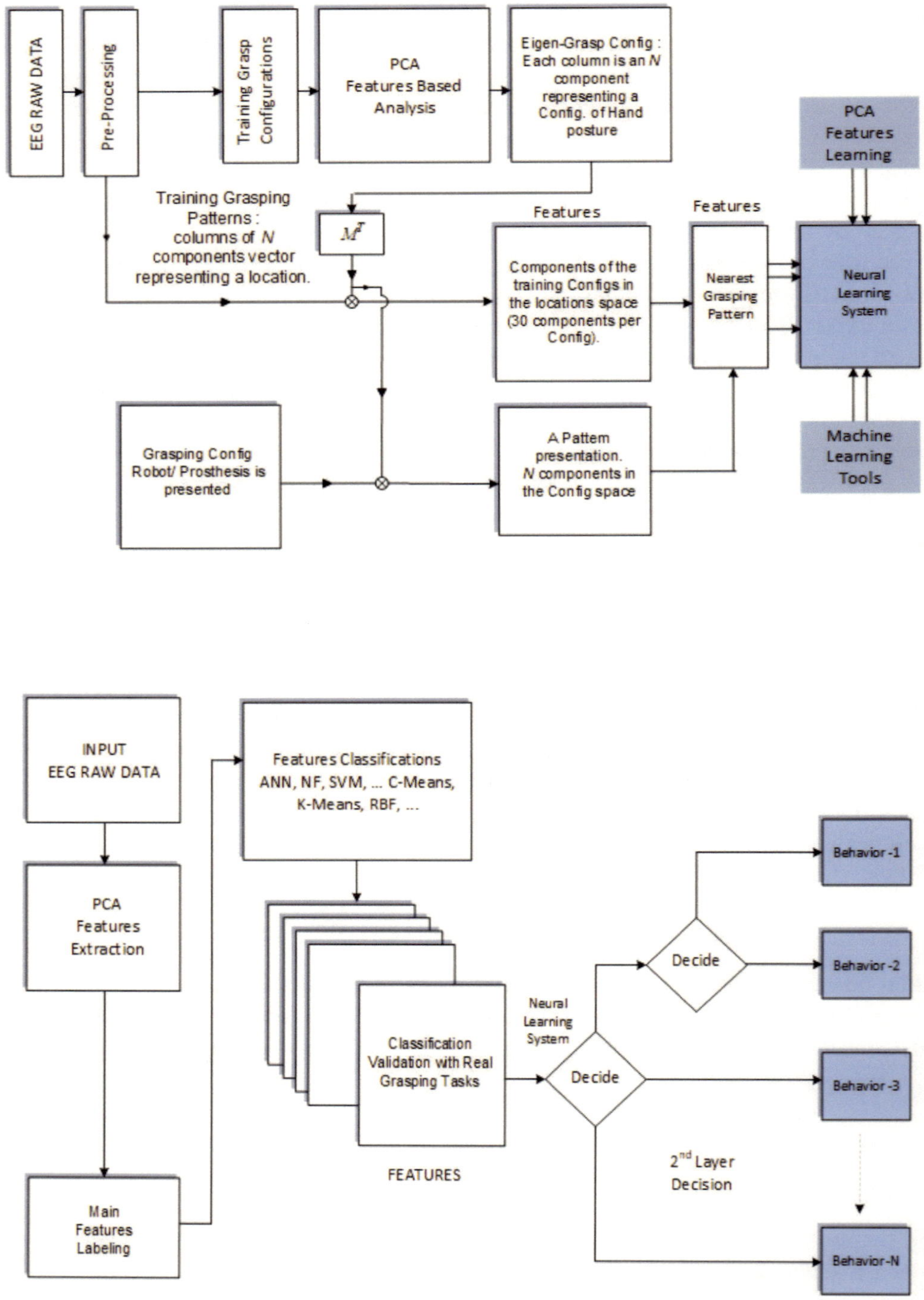

Figure 2. (*Top*): Overall system EEG-PCA detection of main components hierarchy. (*Down*): Layout of how the machine learning tools and techniques can be integrated with the detection process.

(*i*) A human grasping experimental task is initially conducted. The experiment was related to grasping of an object to perform a defined grasping task. Further details about source, and the EEG dataset will be further explained at later stage here in this chapter. Due to raw nature of the detected EEG waves, EEG brainwaves have been filtered, hence they are processed to be much easy to compute and analysis. (*ii*) To identify related events during the defined grasping task for different participates. In his respect, a PCA and fuzzy clustering were applied to the EEG dataset to identify the most identifiable wave characterization. Identified features are then used for copying a human behavioral motion as used for robotic hand.

1.3.2. *Chapter organization*

To achieve the stated objectives, the chapter has been structured as follow. A set of experiments were conducted with advanced EEG detection probes placed over the subject skull, as in Luciw et al. [31]. PCA has been used to detect data features, hence C-fuzzy clustering, is used to learn inherent characterization. Characterization will be further learned to a robotic system.

2. Gasping events detection

2.1. EEG gasping clustering for events detection

In order to detect different events during grasping or manipulations tasks, it is required to classify the data via clustering of resulting EEG waves. This mean to classify different performances. Based on above definitions, fuzzy clustering of the massive EEG dataset is precisely formulated as an optimization problem. This is to minimize the distances to which the different sampled EEG waves are belonging. Formulation of fuzzy clustering is as a minimization problem, i.e. to minimize distances clusters centers u_{ij}:

$$J(u_{ij}, v_k) = \sum_{i=1}^{c}\sum_{j=1}^{n}\sum_{k=1}^{c} g(w(x_i), u_{ij}) d(x_j, v_k) \quad i,k = (1\ 2\ \cdots\ c)\ j = (1\ \ 2\ \ \cdots\ \ n) \tag{1}$$

subject to the following clusters centers constraints:

$$\sum_{i=1}^{c} u_{ij} = 1\ j = 1\ \ 2\ \ \cdots\ \ n \qquad 0 \leq \sum_{j=1}^{n} u_{ij} \leq n\ \ i = (1\ \ 2\ \ \cdots\ \ c) \tag{2}$$

One of the widely employed clustering methods based on Eq. (1) is the fuzzy *c*-means (FCM) algorithm. The objective function of the FCM algorithm is expressed in the form of,

$$J(u_{ij}, v_k) = \sum_{i=1}^{c}\sum_{j=1}^{n} u_{ij}^m \left\| x_j - v_i \right\|^2, \qquad m > 1 \tag{3}$$

In Eq. (3), m is known as exponential weight that influences the degree of fuzziness of the membership or partition matrix. To solve this minimization problem, the objective function

$J(u_{ij}, v_k)$ is differentiated in Eq. (3) with respect to v_i (for fixed u_{ij}, $i = 1,\ldots,c$, $j = 1,\ldots,n$) and to u_{ij} (for fixed v_i, $i = 1,\ldots,c$). Applying constraints conditions, thus obtaining:

$$v_i = \frac{1}{\sum_{j=1}^{n} (u_{ij})^m} \sum_{j=1}^{n} (u_{ij})^m x_j, \qquad i = 1,2,\ldots,c \tag{4}$$

$$u_{ij} = \frac{\left(1/\|x_j - v_i\|^2\right)^{1/m-1}}{\sum_{k=1}^{c} \left(1/\|x_j - v_k\|^2\right)^{1/m-1}}, \qquad i = 1,2,\ldots,c; \quad j = 1,2,\ldots,n \tag{5}$$

The system described by Eqs. (4) and (5) cannot be solved analytically. However, fuzzy c-means algorithm provides a computationally iterative approach to approximating the minimum of the objective function starting from a given position.

2.2. EEG gasping events: dimensionally reduction

2.2.1. Concept of Eigen EEG waves

Principal components analysis (PCA), aims to find total variation in the set of the used EEG waves, hence it helps to explain this variation by less number of variables. PCA does this by computing the basis of a space which is represented by EEG waves. The basis vectors computed by PCA are in direction of the largest variance of used EEG waves. These basis vectors are computed by solution of an eigen problem, and as such the basis vectors are eigenvectors.

In order to deal with the massive dataset resulting from the grasping, we need also to reduce the dimensionality of the data size, as most of the resulting brain waves look similar with very minor differences. This is summarized as follows: Finding the mean of the signals, hence compute the eigenvectors. First we need to find the mean of all the N columns separately by:

$$\overline{x}_n = \frac{1}{M} \sum_{m=1}^{M} x_{mn}, \; n = 1, 2, \ldots, N \tag{6}$$

In Eq. (6), $\overline{x}_n$ is the mean of each N column ($n = 1, 2, \ldots, N$).

The next step is to do cantering the massive EEG waves. Since PCA requires the matrix to be cantered by subtracting the mean from each column. This will cause the mean of each column to be zero.

$$\Delta_n = x_n - \overline{x}_n \tag{7}$$

where Δ_n is the column of the X matrix after subtracting the mean from the original column. This will cause the data to be moved close to the centre (*origin*) of the principal components. We then need also to compute the covariance matrix of the resulting EEG waves. It is required to get the covariance matrix which is the variance between all the columns to get the relationship between these columns in terms of the variance (multiple dimension variance). The

http://dx.doi.org/10.5772/intechopen.72455

variance is how much the data varies from its mean and the covariance is used to find some kind of a relationship between only two dimensions. For example, relationship between velocity, and car crashes. Therefore, the covariance matrix is just all the combinations of the covariance between each dimension and another. The value of the covariance determines the relationship between the two dimensions. If it is positive, then if one dimension increases the other will increase as well. If it is negative, then the relationship is inversely proportional and when one increase the other decreases. Finally, if it is zero, then the two dimensions are independent of each other or have a nonlinear relationship. The magnitude of the covariance will determine the amount of increase that will occur in the other dimension with maximum relationship of one to one. The covariance matrix:

$$\boldsymbol{C} = \frac{\mathbf{1}}{\boldsymbol{N}}\sum_{n=1}^{N} \Delta_n * \Delta_n^T \tag{8}$$

In Eq. (8), $\boldsymbol{\Delta}_n^T$ is the transpose column of $\boldsymbol{\Delta}_n$. This is further expanded into:

$$\boldsymbol{C} = \begin{bmatrix} x_{1,1}-\overline{x}_1 & \cdots & x_{1,N}-\overline{x}_N \\ \vdots & \ddots & \vdots \\ x_{M,1}-\overline{x}_1 & \cdots & x_{M,N}-\overline{x}_N \end{bmatrix} * \begin{bmatrix} x_{1,1}-\overline{x}_1 & \cdots & x_{M,1}-\overline{x}_1 \\ \vdots & \ddots & \vdots \\ x_{1,N}-\overline{x}_N & \cdots & x_{M,N}-\overline{x}_N \end{bmatrix} \tag{9}$$

Finally, the covariance matrix $\boldsymbol{C}$ is further expressed by Subasi and Gursoy [6]:

$$\boldsymbol{C} = \begin{bmatrix} \mathrm{cov}(x_1, x_1) & \cdots & \mathrm{cov}(x_1, x_N) \\ \vdots & \ddots & \vdots \\ \mathrm{cov}(x_N, x_1) & \cdots & \mathrm{cov}(x_N, x_N) \end{bmatrix} \tag{10}$$

The $\boldsymbol{C}$, which is a square matrix, is said to be symmetric around the main diagonal because it was results of multiplication of a matrix and its transpose. The main diagonal of $\boldsymbol{C}$, is the covariance between the dimension and it self. Therefore, once we are looking to find the relationship between the dimensions, we will look at the non-diagonal elements and judge based on their value, as in Subasi and Gursoy [26].

Finally, we need to compute the eigenvalues and eigenvectors. In this respect, to understand patterns of the normalized data, we need to understand the covariance $\boldsymbol{C}$ of the data. This can be possible by getting some lines in the plot that will explain the data pattern based on the covariance of the signals. This can be achieved by getting the eigenvectors of the covariance matrix. We can find eigenvectors for the covariance matrix $\boldsymbol{C}$, this is because of the square nature which is a requirement for eigenvector calculations, Subasi and Gursoy [26]. For an $(n{\times}n)$ matrix A, if we find a row vector X $(n{\times}1)$ that could be multiplied by A and get the same vector X multiplied by a value λ called an eigenvalue and the vector is an eigenvector. Since the matrix A transforms the vector X to scale positions by an amount equal to λ, it is called a transformation matrix as presented by:

$$AX = \lambda X \tag{11}$$

There are (n) eigenvalues for an ($n \times n$) transformation matrix. Since every eigenvector is scaled by an eigenvalue (λ), we shall have (n) eigenvectors as well. The eigenvalues (λ) is found by solving for while using the identity given by equation Eq. (12):

$$(A - I\lambda)X = 0 \tag{12}$$

In Eq. (12), I is an identity/unity matrix which does not alter value of a matrix it's multiple with. Getting the determinant of $|(A-I\lambda)|$, and solving for λ, results in finding eigenvalues. Substituting each (λ) in Eq. (12), and solving for (X), results in the eigenvector (X) for that λ.

$$\det(A - I\lambda) = 0 \tag{13}$$

3. Time-frequency analysis of grasping EEG waves

3.1. Time-frequency relation

It is essential to investigate both time and frequency domain relation among the detected EEG waves. These relations reveal where (in term of frequency), and when (in term of time), and when power densities are taking place during a course of thinking of grasping. EEGLAB in addition to other tools have been used to analyze the brain data both in time and frequency domain. EEGLAB is a Matlab based toolbox that was developed to further analyze the EEG waves both in time and frequency domains, as found in Delorme and Makeig [32]. This shall further observe results of using EEGLAB at different relations among the waves both time and frequency domains. The experiments were conducted by 11 participates. Further analysis on using the EEGLAB are presented at later stages. The grasping experiment can be further found as in reference to Luciw et al. [31]. The entire dataset (11 individuals grasping experiment), was passed into both a PCA algorithm, and fuzzy based clustering algorithm. PAC algorithm was used to reduce the dimensionality of such massive dataset. Fuzzy c-means clustering was used to identify a number of clusters, thus similar dataset events. It is well known, human do similar actions while doing a similar grasping task. We shall identify (P_1), as the first individual that performed the grasping task, whereas (P_9), for example is the ninth individual performing the experiment. Similar naming are given to other individuals.

3.1.1. Time-domain analysis of grasping waves

A) EXAMPLE, FIRST PARTICIPANT: First we loaded dataset for the first person (P_1), and then plotted all channels in time domain. Signals are time stamped with labels of what they correspond to in action that happed in the experiment and was recorded by all of the sensors. These are just the main events and they include the LED turning on and off, when the hand starts moving, when each finger touches the plates, the object lift off the table and replacing it back to original position, if the new trial includes expected or unexpected high or low weights, and finally the release of the fingers from the plates. The all the possible labels are clearly shown in **Table 1**. In **Figures 3** and **4**, after the LED was turned on there was a minor increase

http://dx.doi.org/10.5772/intechopen.72455

Event no.		Label	Meaning
Event-01	–	tTouch	Fingers touching
Event-02	–	tStartLoadPhase	Stating to apply load
Event-03	–	tReplace	Replacing the object
Event-04	–	tRelease	Releasing fingers
Event-05	___	tLiftOff	Lifting the object
Event-06	___	tHandStart	Hand movement started
Event-07	___	Unexp.light weight lift	Unexpected light weight
Event-08	___	Unexp.heavy weight lift	Unexpected heavy weight
Event-09	___	LEDON	LED turned On
Event-10	___	LEDOFF	LED turned Off
Event-11	___	Expected Weight Lift	Expected weight

Table 1. Grasping events labels, (Meaning for EEGLAB plot).

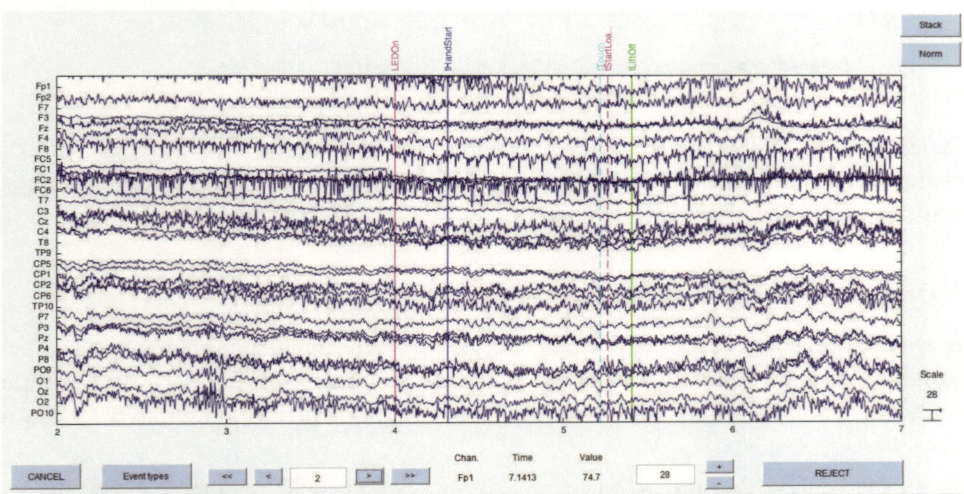

Figure 3. EEGLAB time-domain plot for (P_1), and all 32 channels (*Voltage* (μV) *vs. Time* (*Sec.*)).

in the voltage due to the intent to move in response to the event (ERP). If the hand started moving there was a minor increase in some channels and decrease in the reference channels due to hand movement. Consequently, as fingers touched the object, forces were applied. Minor change happened until the lift off, which caused a high-low voltage for lift and then relax in destination after (0.5 sec) from lifting off. In anticipation of LED turning on, there was an increase in voltage and a sudden spike once LED tuned off due to eye blinking. Finally, after releasing the object, there was a drop in voltage when relaxing the fingers and returning into original position.

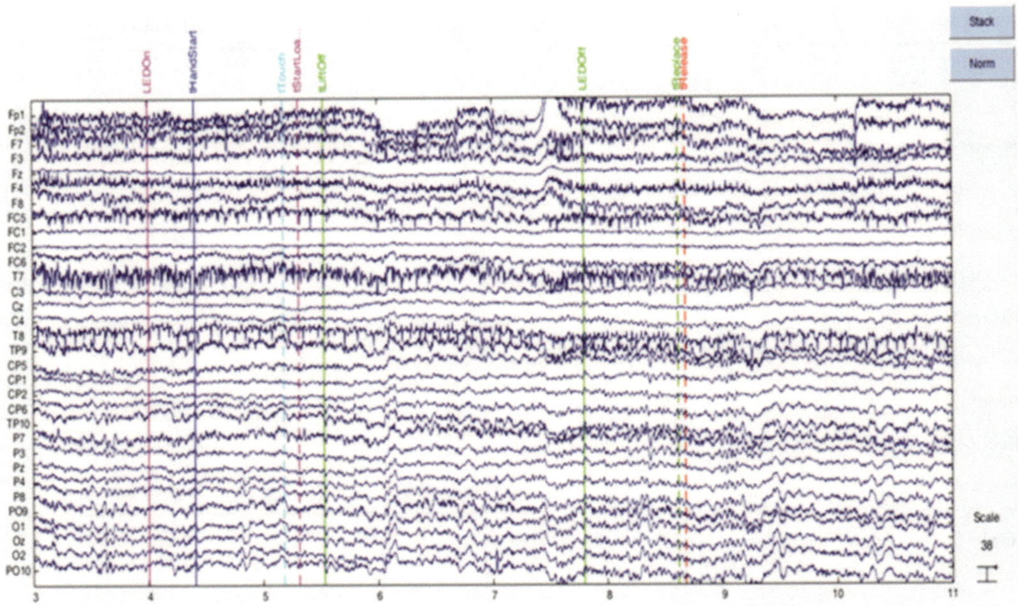

Figure 4. EEGLAB time-domain plot for (P_4), and all 32 channels (voltage (μV) *vs. Time* (*Sec.*)).

B) EXAMPLE, SECOND PARTICIPANT: Second, is the plot for the second person (called (P_4) in the data) in a similar trial as the first person. This looks almost exact and similar as the first participant, except the blinking happened just before the LED turning off.

3.1.2. Frequency- analysis of grasping waves: power spectrum

After we are done with the time domain analysis, we will now examine the spectral power changes and the corresponding area of the brain that all the electrodes cover. This will indicate any movement and its origin within the human head. This is further shown in both **Figures 5** and **6**.

We should indicate that, while the data were recorded by (32) channel brain waves cap, such massive dataset are plotted in **Figure 7**. **Figure 7** shows the time plot of only nine out of 32 channels that were used during the conduct of the experiment. The PCA and clustering operations were performed over the entire (11 individuals – nine trails – different objects – different weights), hence clustering operations were performed. This is further shown both in **Figure 8**.

FIRST PARTICIPANT, SECOND PARTICIPANT: In reference to the frequency spectrum during the grasping tasks, while looking at **Figure 5** we can deduce that power changes accrued at (5.9 Hz) in theta band and corresponded to the frontal part of the brain which indicates relaxation as shown in previous studies. There were only minor changes at (9.8 Hz) in alpha band. Finally, there were big power changes in (F8, FC6, and T8) and minor to the counterpart

http://dx.doi.org/10.5772/intechopen.72455

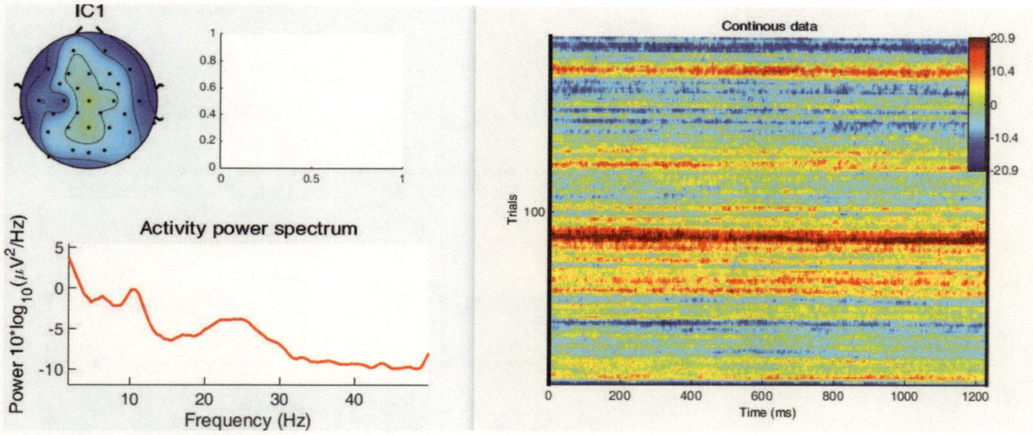

Figure 5. EEGLAB frequency-domain power spectrum analysis, for (P_1) for all (32 channels). ICA for the resulting grasping waves.

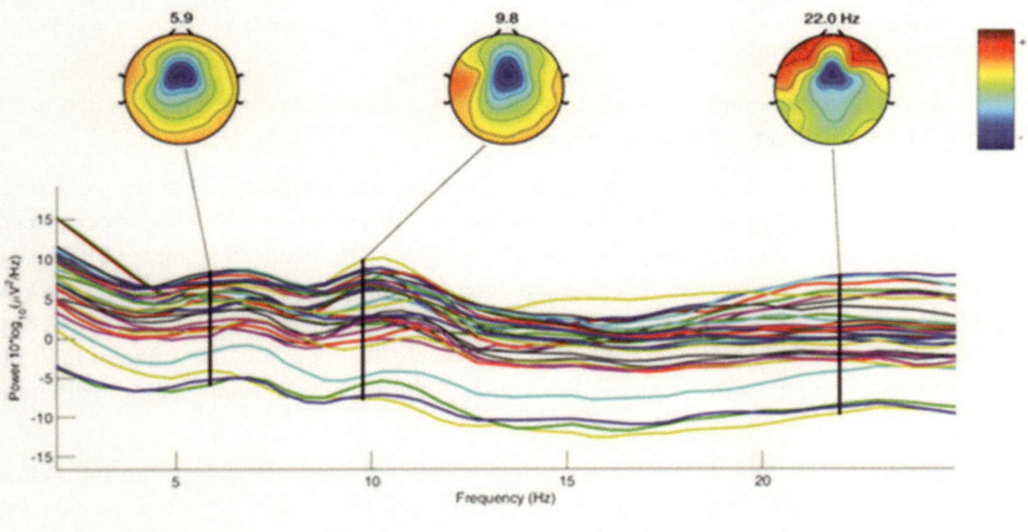

Figure 6. EEGLAB frequency-domain power spectrum for the grasping trails, for (P_4) for all (32 channels).

at (22 Hz) in beta band side this could indicate the blinking artifact we saw earlier. We can see similar patterns in **Figure 6** for the second person especially for blinking artifact in (22 Hz) in beta band. However, there is significant difference in at (9.8 Hz) which could be due to the finger or hand movement which will be investigated in the next section using PCA. For each real grasping experiment, it was found the identical and similar EEG patterns, that was detected by the locations of the clustered centers and gathered fuzzy membership function. In this context, the gathered fuzzy memberships obtained through the fuzzy c-mean clustered, do indicate the inherent knowledge about how the grasping was conducted. This knowledge is

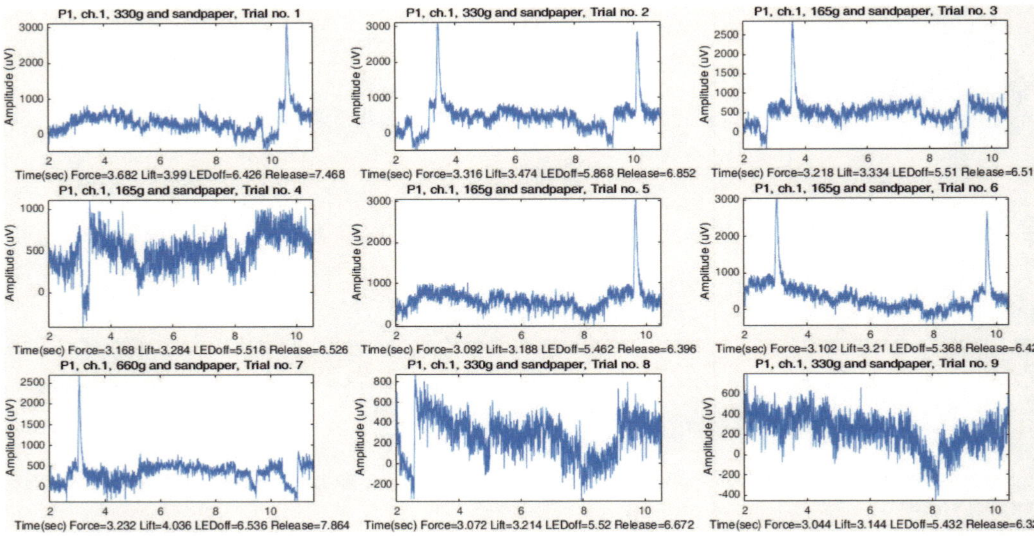

Figure 7. Plot of (P_1) for (9 trials) by the individuals. Note: here the weight changes but the type is the same (sandpaper). Full time, same all-channels, and (9) trials (only weight change). (P_1), Full time, all-channels, and (9) trials (series 1). (P_1), Full time, all-channels, and (9) trials.

further decoded for generating the most suitable patterns of motorizing finger motion to be used for the robotic hand, or the human type prosthesis.

In reference to **Table 2**, the adopted dimensionality reduction via PCA, hence the clustering, have helped in identifying there main events related to this experiment which was performed during 6 seconds of time. Detecting EEG similar events during hand grasping is not an easy task to achieve. Such similar events do indicate how the real human behavior was done. Once this is done, this can be copied to a robotic system. Combining both PCA with clustering mechanism has definitely help the inherent and hidden features on how the different individuals performs grasping. PCA has proven the similar patterns, despite of the shapes of different waves.

The inherent features are almost the same, as seen via the eigenvalues of the individual experiments and the individual personals. In addition, fuzzy clustering has proven the overlapped shapes of the EEG waves during the same experiments, however, despite of such overlaps, the fuzzy clusters are able to detect the different grasping events that can be translated to a robotic hand. Within this chapter, we have presented a mechanism through which to analyze brainwaves EEG patterns for robotics and prosthesis use. The main discussed theme was based on using pattern recognition and clustering techniques for capturing the inherent patterns characterization during a phase of grasping. Capturing events and these characterizations are useful for transferring human knowledge (for the grasping tasks) to robotics hand, or prosthetics for complex grasping. In achieving that, a fuzzy clustering technique was able to identify similar events. The presented methodology was very useful in getting the intrinsic patterns. The presented approach also is efficient for mapping useful human thoughts, for robotics grasping tasks.

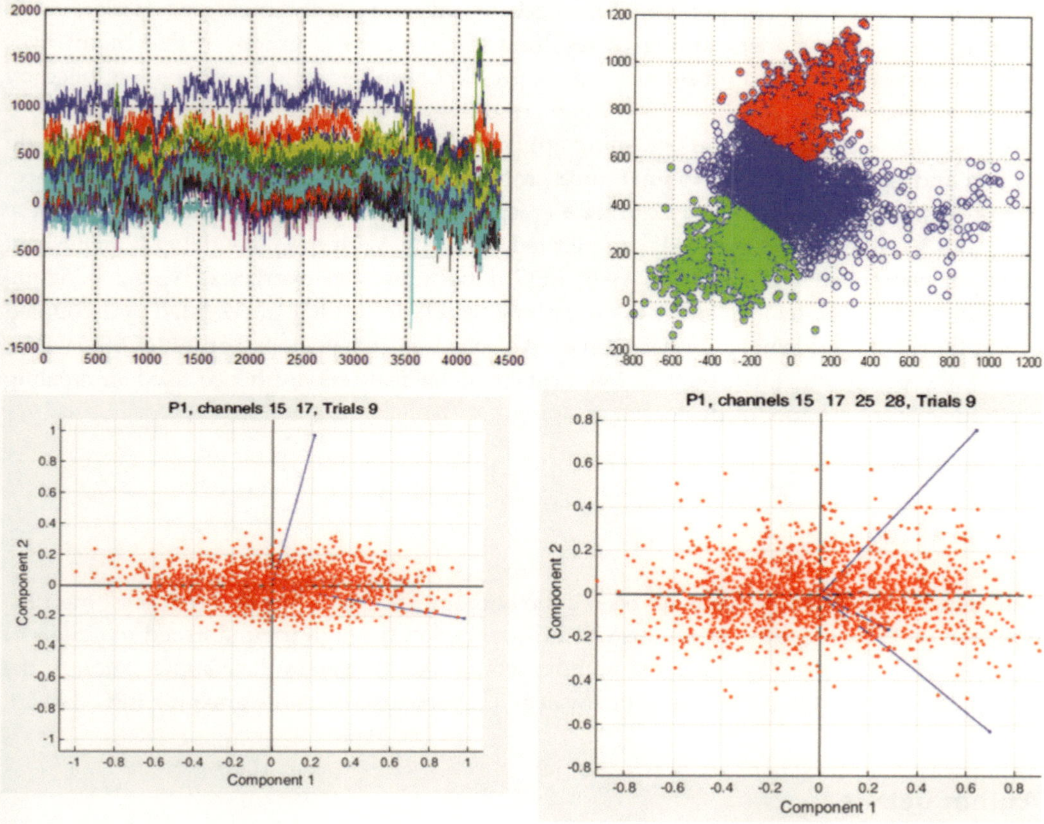

Figure 8. Three regions of clustering patterns for second experimentation. Grasping EEG waves as being clustered into three main clusters. Each cluster indicates the focus of where similar EEG patterns are centered. *Above-Left*: Time domain signals during the course of grasping. *Above-Right*: Corresponding c-means clustering results. Four to Five main clusters have been identified for all the trails. Down-*Left-Right*: PCA analysis of the resulting EEG waves during the grasping task. Creating principle components does lead to observe similar waves, thus reducing the dimensionally of the EEG grasping data size.

	TASK 1	Event-1	Event-1	Overlapped Clusters
1	Hand motion	Hand grasping	Hand grasping	Clusters 1,3
2	Figures motion	Apply force	Apply force	Clusters 2,3
3	Hand motion	Finger move	Force releasing	Clusters 3,2

Table 2. Grasping experimentation and patterns mapping and detected fuzzy clusters association.

4. Conclusion

Electroencephalography and brain waves for robotics, human, computer, and machine interfaces (BMI, BCI), are evolving very fast. This is due to the fast development of high tech interfacing devices and computational methods. In this respect, the presented work is dedicated towards deep understanding of resulting electroencephalography (EEG) brainwaves during a typical grasp and lift human grasping task. During grasping, forces are applied by fingertips dexterously, as observed through resulting EEG waves. For mirroring this to a prosthesis-robotics and dextrous grasping applications, methods have to be developed to do mining and catch features for optimal forces, movements, and accurate finger joints displacements. Resulting EEG brainwaves during a grasp and lift task are very useful, however these EEG waves are related, correlated, complicated, and raw. With the potential and analysis of principal components analysis (PCA) of EEG, it indicated an overlap of valuable neural behaviors from various locations over the human skull, indicating interrelated and coupled events for robotic grasping. PCA has been used also to unlock few main features of EEG waves during the grasp and lift task. The foremost grasping features are hence used in creating accurate events for a robotic dexterous grasping.

Acknowledgements

This research work would have not been done without using accurate and real grasping EEG brainwaves. Grasping dataset was used with permission through the published materials of Luciw et al. [31]. Given this fact, the authors would like to express their appreciation to the Ume University, and in particular to Luciw et al. [31], for supplying the grasping EEG dataset.

Author details

Ebrahim A. Mattar[1]*, Hessa J. Al-Junaid[2] and Hamad H. Al-Seddiqi[2]

*Address all correspondence to: ebmattar@uob.edu.bh

1 College of Engineering, University of Bahrain, Sukhair, Kingdom of Bahrain

2 College of Information Technology, University of Bahrain, Sukhair, Kingdom of Bahrain

References

[1] Vincent V et al. Biomimetics: its Practice and Theory. 22nd August 2006. DOI: 10.1098/rsif.2006.0127

[2] Andrew P, Harshavardhan A, José L. Contreras-Vidal: Decoding repetitive finger movements with brain activity acquired vi anon invasive electroencephalograph. Frontiers in Neuroengineering; 13th March 2014. DOI: 10.3389/fneng.2014.00003

[3] Zaepffel M, Trachel R, Bjørg Elisabeth K, Brochier T. Modulations of EEG beta power during planning and execution of grasping movements. Plos One. March 2013;**8**(3):e60060

[4] Sam B, Schwartz R, Holmes D. An EEG-based Brain Computer Interface for Rehabilitation and Restoration of Hand Control following Stroke Using Ipsilateral Cortical Physiology. 2014 Report, Washington University

[5] Theodore B, Robert H, Dong S, Anushka G, Vasilis M, and Deadwyler S. A cortical neural prosthesis for restoring and enhancing memory. Journal of Neural Engineering August. 2011;**8**(4):046017. DOI:10.1088/1741-2560/8/4/046017

[6] Yuanfang R, Yan W, Yanbin G. A co-training algorithm for EEG classification with biomimetic pattern recognition and sparse representation. Journal of Neurocomputing. 2014;**137**: 212-222

[7] Perruchoud D, Pisotta J, Carda S, Murray M, Ionta S. Biomimetic rehabilitation engineering: The importance of somatosensory feedback for brain–machine interfaces. Journal of Neural Engineering. 2016;**13**, 041001:9

[8] Menniti D, Pullano S, Bianco M, Citraro R, Russo E, De Sarro G, Fiorillo A. Biomimetic Sonar for Electrical Activation of the Auditory Pathway. Hindawi Journal of Sensors. 2017; **2017**. Article ID 2632178

[9] Bullock I, Zheng J, Rosa S, Guertler C, Dollar A. Grasp frequency and usage in daily household and machine shop tasks. IEEE Transactions on Haptics. 2013;**6**(3):296-308

[10] Bullock M, Feix T, Dollar M. Finding small, versatile sets of human grasps to span common objects. Robotics and Automation (ICRA), IEEE International Conference on, Karlsruhe, 6–10 May 2013, 2013. pp. 1068-1075

[11] Bullock M, Feix T, Dollar M. The Yale human grasping dataset: Grasp, object, and task data in household and machine shop environments. The International Journal of Robotics Research. 2014;**34**(3):251-255

[12] Feix T, Bullock L, Dollar A. Analysis of human grasping behavior: Object characteristics and grasp type. IEEE Transactions on Haptics. 2014;**7**(3):311-323

[13] Khorshidtalab A, Salami M, Hamedi M. Robust classification of motor imagery EEG signals using statistical time–domain features. Journal of Physiol Measurements. 2013;**34**: 1563-1579

[14] Mao X, Mengfan L, Wei L, Niu L, Xian B, Zeng M, Chen G. Progress in EEG-based brain robot interaction systems. Hindawi Computational Intelligence and Neuroscience Journal. 2017;**2017**:1-25. Article ID 1742862. https://doi.org/10.1155/2017/1742862

[15] Feix T, Bullock I, Dollar A. Analysis of human grasping behavior: Correlating tasks, objects and grasps. IEEE Transactions on Haptics. 2014;**7**(4):430-441

[16] Subasi A, Ismail Gursoy M. EEG signal classification using PCA, ICA, LDA and support vector machines. Journal of Expert Systems with Applications. 2010;**37**:8659-8666

[17] Al-Qazzaz N, Bin Mohd Ali S, Anom S, Islam A, Escudero J. Automatic artifact removal in EEG of normal and demented individuals using ICA–WT during working memory tasks. Journal of Sensors. 2017;**17**:1326. DOI: 10.3390/s17061326

[18] Yoshioka M, Zhu C, Imamura K, Wang F, Yu H, Duan F, Yan Y. Experimental design and signal selection for construction of a robot control system based on EEG signals. 2014, in Yoshioka et al. Robotics and Biomimetics. 2014;**1**:22

[19] Hazrati M, Erfanian A. An online EEG-based brain–computer interface for controlling hand grasp using an adaptive probabilistic neural network. Journal of Medical Engineering and Physics. 2010;**32**(7):730-739

[20] Jianjun M, Shuying Z, Angeliki B, Jaron O, Bryan B, Bin H. Noninvasive electroencephalogram based control of a robotic arm for reach and grasp tasks. Scientific Reports. 14th of December 2016;**6**:38565. DOI: 10.1038/srep38565

[21] Vinet R, Lozac'h Y, Beaundry N, Drouin G. Design methodology for a multifunctional hand prosthesis. Journal of Rehabilitation Research and Development. 1995;**32**:316-324

[22] Cutkosky M. Robotic Grasping and Fine Manipulation. Boston: Kluwer Academic Publishers; 1985

[23] Mattar E, Al-Junaid H. Fuzzy C-Means Classification of Electroencephalography (EEG) Waves for Robotic System Time Events and Control. Conference: 2016, the 14th Pacific Rim International Conference on Artificial Intelligence, at *PRICAI 2016*, Thailand, vol. 1, ISBN 978-616-92700-1-0, pp. 22-26

[24] Dustin J. Tyler: Creating a Prosthetic Hand That Can Feel. Publication of the IEEE Spectrum. April 2016

[25] Touch Bionics: A Story of Innovation and Growth Powered by Co-investment. Available from: http://www.cando.scot/case-studies/touch-bionics/. [Accessed: 2017-11-10]

[26] Subasi A, Gursoy M. EEG signal classification using PCA, ICA, LDA and support vector machines. Expert Systems with Applications. 2010;**37**(12):8659-8666

[27] Wang C, Zou J, Zhang J, Wang M, Wang R. Feature extraction and recognition of epileptiform activity in EEG by combining PCA with ApEn. Cognitive Neurodynamics. 2010;**4**(3): 233-240

[28] Mahajan K, Vargantwar M, Rajput S. Classification of EEG using PCA, ICA and neural network. International Journal of Engineering and Advanced Technology. 2011;**1**(1):80-83

[29] Aminian F. Electroencephalogram (EEG) signal classification using neural networks with wavelet packet analysis, principal component analysis and data normalization as preprocessors. In: Vrajitoru D, editor. Proceedings of the Twenty-First MAICS 2010 Midwest Artificial Intelligence and Cognitive Science Conference (pp. 55-62). South Bend, IN: Midwest Artificial Intelligence and Cognitive Science Conference; 2010

[30] Jin J, Wang X, Wang B. Classification of Direction Perception EEG Based on PCA-SVM. 2007, Natural Computation. ICNC 2007. 3rd International Conference on, Haikou, pp. 116-120

[31] Luciw M, Jarocka E, Edin B. Multi-channel EEG recordings during 3,936 grasp and lift trials with varying weight and friction. Scientific Data. 2014, article number: 140047. pp. 1-11. DOI: 10.1038/sdata, 2014. 47

[32] Delorme A, Makeig S. EEGLAB: An open source toolbox for analysis of single-trial EEG dynamics including independent component analysis. Elsevier, Journal of Neuroscience Methods. 2004;**134**:9-21

Chapter 3

Sensory Feedback Device for Myoelectric Prosthetic Hand

Chiharu Ishii

Additional information is available at the end of the chapter

http://dx.doi.org/10.5772/intechopen.71690

Abstract

In this work, a sensory feedback device for myoelectric prosthetic hand was developed to enhance the quality of life (QOL) of myoelectric prosthetic hand users. Two types of sensory feedback, namely, force sense feedback and temperature sense feedback, were proposed. As for the feedback device of force sense, the device is mounted on the user's upper arm and provides hardness of the object onto the upper arm by winding a belt using a motor. On the other hand, as for the feedback device of temperature sense, the device is mounted on the user's upper arm and presents temperature of the object onto the upper arm using a Peltier element. Finally, two-sensory feedback devices were united, and a two-sensory feedback device was built.

Keywords: sensory feedback device, myoelectric prosthetic hand, force sense, temperature sense

1. Introduction

A myoelectric prosthetic hand is an electrically driven artificial hand that is controlled based on biosignals generated by muscle movement. Therefore, the myoelectric prosthetic hand can be moved freely as intended by the user. However, the user cannot feel sensations when he/she touches an object using the prosthetic hand. Healthy person uses the tactile sense and temperature sense to check the state of the object that he/she touched. A myoelectric prosthetic hand cannot produce any sensations. Therefore, the user has to operate the prosthetic hand only based on visual information. Thus, the user needs to watch the object constantly. This is a burden to the user. To solve this problem, sensory feedback which provides sensation to the user has been studied. Sensory feedback systems for an upper limb prosthesis in the initial stage have been reported in [1]. Summary of studies involving sensory feedback in upper limb prosthetics is listed in [2].

A method of providing a sense of touch realized by vibration stimulation was proposed in [3], and a method of sensory feedback realized by the combination of vibration stimulation and electric stimulation was proposed in [4], respectively. On the other hand, Otsuka et al. [5] developed a device that perceives the temperature when an object was touched by a myoelectric prosthetic hand using a hot and cold pad. Morimitsu and Katsura [6] examined transfer of temperature sense using the Peltier element.

In this study, two types of sensory feedback device were developed to enhance the quality of life (QOL) of myoelectric prosthetic hand users. First, a compact feedback device of force sense (hereinafter referred to as FFB device) with a safety mechanism was developed. The FFB device is mounted on the user's upper arm, and when a prosthetic hand holds an object, a belt in the device is winded by a motor to present the holding force to the user's arm. Besides, the winding speed of the belt is changed according to the hardness of the object held by the prosthetic hand. In the control system of the FFB device, a reference input creation model creates reference input signals according to the hardness. A self-tuning proportional-integral-derivative (PID) control method proposed by [7] was employed to adjust the gain of the PID controller based on the state of a target object and control the belt-winding speed of the FFB device following the reference input. For the verification of the effectiveness of the control system for the FFB device, a myoelectric prosthetic hand made by [8] was combined with a control method proposed by [9], and experiments to distinguish among five kinds of springs of different hardness were conducted.

Second, a feedback device of temperature sense (hereinafter referred to as TFB device) was developed by using a Peltier element to present an object temperature when the user touches the object by a prosthetic hand. A temperature prediction algorithm was proposed to shorten the temperature measurement. Besides, the temperature sense differs at each body site. Therefore, the TFB device developed in this study transfers temperature sense felt by the fingertip to the upper arm based on the result of the experiment on temperature sense investigation. Furthermore, experiments to distinguish among five different temperatures were performed to verify the effectiveness of the TFB device. Finally, two-sensory feedback devices were united, and a two-sensory feedback device was built.

2. Myoelectric prosthetic hand

In this study, the myoelectric prosthetic hand made by [8] shown in **Figure 1(a)** is used. The prosthetic hand consists of motors and wires, and the fingers are bent by winding the wires. A pressure sensor is attached to a finger cushion on the prosthetic hand's index finger, and a temperature sensor is attached at the fingertip of the prosthetic hand's middle finger.

The myoelectric prosthetic hand used in this study only has three fingers, namely, the thumb, index finger, and middle finger. Therefore, the prosthetic hand grasps an object by bending

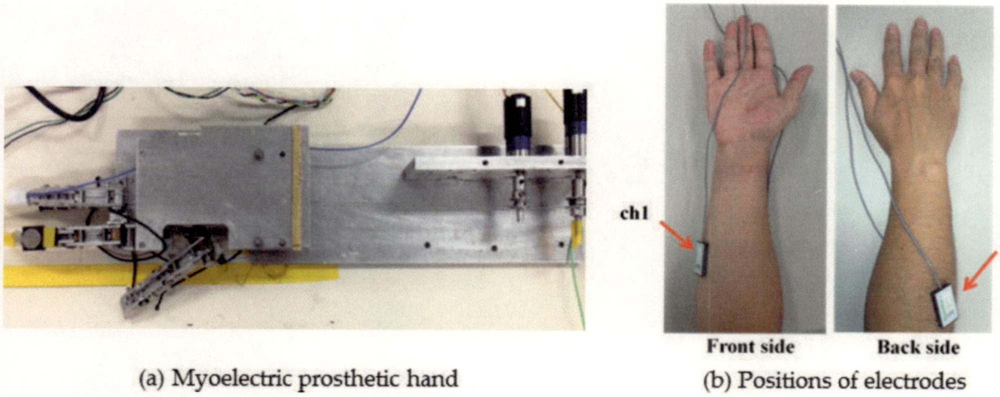

Figure 1. (a) Myoelectric prosthetic hand. (b) positions of electrodes.

the index finger for contact force feedback case and bending the middle finger for temperature sense feedback case. The motion of the index finger and the middle finger is identified by measuring the surface electromyogram (SEMG) of the flexor digitorum superficialis (ch 1) and extensor carpi radialis longus (ch 2) shown in **Figure 1(b)** to control the proximal interphalangeal (PIP) joint of the prosthetic hand's index finger and middle finger.

3. Feedback device of force sense (FFB device)

3.1. Design concept

In this study, amputees who lost the lower half of a single forearm were chosen as subjects, and electrodes were placed on the upper half of the forearm to operate a prosthetic hand. To let prosthetic hand users recognize the sense more intuitively, the sense of pressure was given to the prosthetic hand users when they grasp an object. The sense of pressure was presented by the tightening force of a belt on the upper arm of the users.

The difference of the contact force according to the hardness of the object is expressed as the winding speed of the belt. The pressure value added to the finger and displacement of the finger are measured, and they are used to estimate the object hardness. Then, the winding speed of the belt is changed to present the estimated hardness. Namely, the high winding speed is for a hard object, and the low winding speed is for a soft object.

3.2. Overview of FFB device

3.2.1. Equipment

Figure 2 shows the developed FFB device and its attached state on the user's upper arm. The working principle is described as follows. Contact force between an object and a finger is measured by the pressure sensor attached to the finger cushion on the prosthetic hand's

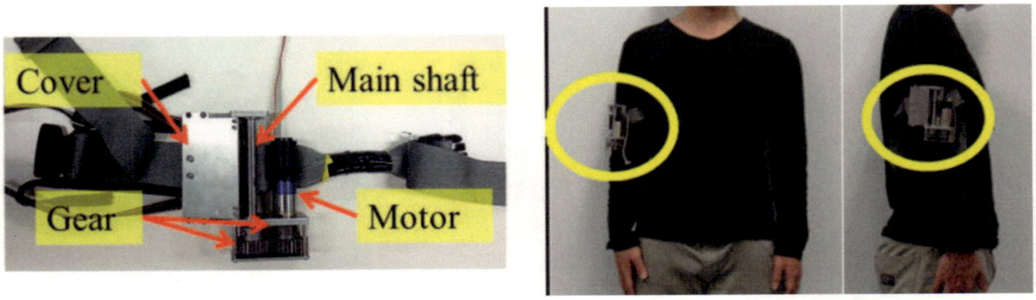

Figure 2. FFB device and its attached state.

index finger. The main shaft for winding the belt and the motor are connected through two gears. When the prosthetic hand grasps the object, namely, when the contact force is sensed, the motor rotates. Therefore, the main shaft is also rotated by the rotation of the motor. Thus, the belt is winded and tightens the upper arm.

The device is small with the dimensions 97 mm (width), 117 mm (depth), and 39 mm (height).

3.2.2. *Safety mechanism*

A safety measure against the motor's failure or other emergent case needs to be taken. In the case of emergency, the belt is released by simply opening the cover, and then the device is released from the arm.

3.3. Control method of prosthetic hand

In this study, a control strategy proposed in [9] is used for the operation method of the myoelectric prosthetic hand. It is well known that an integrated electromyogram (IEMG) reflects muscle activity. Hence, the IEMG is employed to identify the input motion for the operation of the prosthetic hand, and a support vector machine (SVM), which is one of the techniques of the machine learning, is used as an identifier. For the control of the prosthetic hand, a target angle of the finger of the prosthetic hand is set based on how long the user keeps the muscle force. This allows the user to arbitrarily control the finger angle.

3.4. Measurement of hardness

A pressure sensor "FSR402 Short Tail" made by Interlink Electronics was attached on the fingertip to measure the reaction force F [N] from the grasped object. The displacement of the fingertip [m] is measured from the encoder attached on the driving motor of the finger. Then, the spring constant K [N/m], hereinafter referred to as the hardness parameter, can be calculated from Hooke's law:

$$K = \frac{F}{x} \tag{1}$$

A preliminary experiment was conducted in [10] to derive a conversion formula from the pressure (voltage V [v]) measured with the pressure sensor to the reaction force F [N]. The resulting conversion formula is the following:

$$F = 9.81 \times \{(8.01 \times 10^{-3})\, V^6 - (9.79 \times 10^{-2})\, V^5 + (4.52 \times 10^{-1})\, V^4 - (9.49 \times 10^{-1})\, V^3 + (8.56 \times 10^{-1})\, V^2 - (1.37 \times 10^{-1})\, V\} \quad (2)$$

With Eq. (2), a reaction force is calculated from the measurement value obtained with the pressure sensor. Thus, one can calculate the hardness parameter from Eq. (1).

3.5. Control of FFB device

3.5.1. Configuration of reference input creation model

For the purpose of this study, the winding speed of the belt in the FFB device needs to be adjustable according to the hardness of the grasped object. A reference input creation model is defined as shown in **Figure 3**, in which a reference input $r(t)$ is obtained by a step input $u_s(t)$ passing a primary delay filter.

The following relation is obtained from **Figure 3**:

$$R(s) = \frac{1}{Ts+1} U_s(s) \quad (3)$$

A small time constant of the primary delay filter produces a rapidly rising reference input, and a large time constant produces a gradually rising reference input. The time constant T [s] of the primary delay filter is adjusted in accordance with the hardness parameter, K, to control the belt-winding motor to follow the reference input. Thus, T is denoted as $T(K)$.

For safety reason, the FFB device is configured so that up to 10 mm of the belt is wound up.

3.5.2. Relation between time constant and hardness parameter

A function of the time constant $T(K)$ related to the hardness parameter K was derived in [10], which was determined as follows by trial and error:

$$T(K) = \tan^{-1}(-K \times 0.000734 + 2.1) \times 0.705 + 1.202 \quad (4)$$

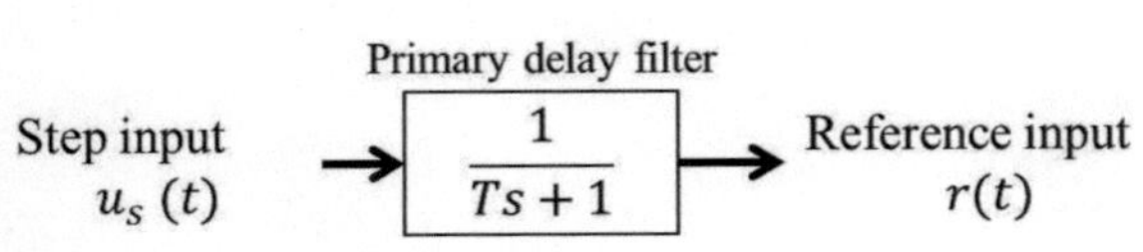

Figure 3. Reference input creation model.

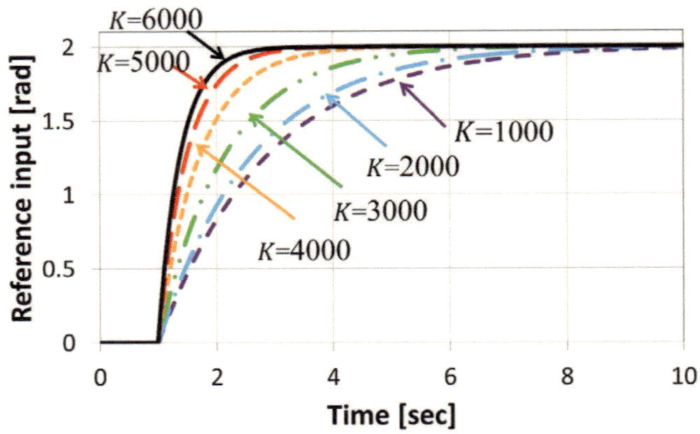

Figure 4. Reference inputs.

3.5.3. Verification of reference input creation model

Numerical simulation was performed to verify the reference input creation model. In the verification, the hardness parameter was increased from 1000 to 6000 for each 1000, and the reference input derived from the reference input creation model was checked. The result is shown in **Figure 4**.

Figure 4 shows that the rise of the curve becomes rapid as the hardness parameter increases. This result indicates that the reference input could be efficiently created so that the user can feel a difference of the hardness.

3.5.4. Self-tuning PID control

The FFB device gives feedback of the force sense by pressing the upper arm of the user. The amount of fat and muscle of the human arm differs from person to person, and the arm hardness could also change depending on how much the user strains the arm. Thus, the arm hardness has a nonlinear characteristic. Therefore, an adaptive control is used for the control of the belt-winding motor of the FFB device. To this end, the self-tuning PID control proposed by [7] is used. Since the self-tuning PID control scheme is based on a discrete time control, Eq. (3) is discretized by a bilinear transformation to design a controller of the FFB device.

The control aim is to determine the control input $u(k)$ so that the output angle $y(k)$ of the FFB device follows the reference input $r(k)$. The detail of the derivation of the control input is described in [10]. The PID gain can be calculated by using the estimation parameters $\hat{a}_1(k)$, $\hat{a}_2(k)$, and $\hat{a}_3(k)$ as follows:

$$\hat{K}_p(k) = \frac{2\hat{a}_2(k) + \hat{a}_3(k) - 2}{\hat{a}_1(k)}, \quad \hat{K}_i(k) = \frac{1}{\hat{a}_1(k)}, \quad \hat{K}_d(k) = \frac{1 - \hat{a}_2(k) - \hat{a}_3(k)}{\hat{a}_1(k)} \tag{5}$$

Here, k is the number of steps. Then, the control input is given by the following equation:

$$u(k) = u(k-1) + \hat{K}_p(k)\{y(k-1) - y(k)\} + \hat{K}_i(k)\{r(k) - y(k)\} + \hat{K}_d(k)\{2y(k-1) - y(k) - y(k-2)\} \quad (6)$$

3.6. Operation check of FFB device

For the verification of the functions of the whole system, the prosthetic hand was operated based on the SEMG of a subject, and the hardness of an object that was grasped by the hand was estimated. Then, how the FFB device controlled by the self-tuning PID could follow the reference input created from the estimated hardness was examined.

Five kinds of springs with different hardness were chosen as objects to grasp. **Table 1** shows the physical properties of the springs.

The subject was an adult male in 20s. The results in which each of the springs was grasped once are shown in the following figures and tables. **Figure 5** shows how the FFB device followed the reference input. **Table 2** shows the results of the hardness estimation. **Table 3** shows self-tuning PID gains and the time constant calculated from the estimated hardness.

Figure 5(a) and **(b)** shows a time delay of about 0.2 s in the response of the FFB device for any spring. However, the device clearly followed the reference input. Although a small error between the estimated value and the real value of the hardness parameter is seen in **Table 2**, the time constant was successfully calculated for all of the springs as shown in **Table 3**. Therefore, it was confirmed that the reference input could be created according to the estimated hardness of the grasped object. In addition, the belt-winding action of the FFB device under the control of the self-tuning PID controller followed the reference input obtained from the estimated hardness. Thus, it was verified that the system worked properly.

3.7. Hardness identification experiment

The usefulness of the FFB device with the proposed system implemented is objectively verified with a psychophysics experiment method. Five kinds of springs, shown in **Table 1** in the previous section, were used as the target of the hardness identification.

No.	Spring constant K [N/m]	Diameter [m]	Length [m]
S_1	990	14×10^{-3}	25×10^{-3}
S_2	1800	8×10^{-3}	25×10^{-3}
S_3	2980	10×10^{-3}	30×10^{-3}
S_4	4390	13×10^{-3}	25×10^{-3}
S_5	5340	7×10^{-3}	25×10^{-3}

Table 1. Physical properties of springs.

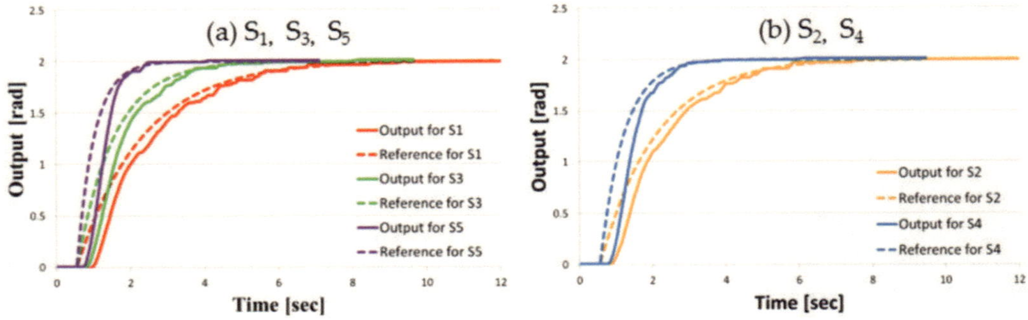

Figure 5. Reference input and output of the FFB device.

Spring	Force [N]	Displacement [m]	Estimated hardness parameter [N/m]	Real hardness parameter [N/m]
S_1	5.23	3.76×10^{-3}	1390	990
S_2	5.27	2.56×10^{-3}	2060	1800
S_3	5.66	1.80×10^{-3}	3150	2980
S_4	5.15	1.27×10^{-3}	4060	4390
S_5	4.96	0.99×10^{-3}	5030	5340

Table 2. Estimated and real hardness parameter.

Spring	Time constant *T* [s]	PID gain		
		$\hat{K}_p(\mathrm{k})$	$\hat{K}_i(\mathrm{k})$	$\hat{K}_d(\mathrm{k})$
S_1	1.78	0.22706	0.0125	0.01154
S_2	1.58	0.22707	0.0125	0.01154
S_3	1.06	0.22713	0.0125	0.01150
S_4	0.69	0.22710	0.0125	0.01151
S_5	0.49	0.22717	0.0125	0.01147

Table 3. Time constant and self-tuned PID gains.

3.7.1. Overview of the experiment

In the experiment, the myoelectric prosthetic hand was not used, but the spring constant of each spring was input directly to a computer, and then the hardness identification experiment using the FFB device was conducted.

The experiment was conducted by following the procedure of a constant method. The hardness of a brass spring ($K = 2980$) was used as standard stimulation, and the hardness of five springs, including the brass spring, was used as comparative stimulation. The experimental procedure is described as follows:

1. FFB device is attached on the upper arm of the subject.
2. The subject is trained so that he can recognize the behavior of the FFB device.
3. The standard stimulation is given to the subject.
4. A 4-second interval is taken.
5. The comparative stimulation is given to the subject.
6. The subject answers which stimulation is harder or whether the two are the same.
7. 25 sets of operations [(3)–(6)] are conducted. In the operations, all the comparative stimulations are used five times in random orders.
8. The operations [(3) and (5)] were replaced, and 25 sets of the operations [(3)–(6)] are conducted again.

The experiments were performed for five healthy subjects in their 20s. In the operations, a 1-min break was taken every five sets to prevent the subject from getting tired.

3.7.2. Results

Table 4 shows the results of experiments. The bold numbers show the ratio of correct identification of the stimulations.

From the results on the hardness identification experiment, one can see that the proposed method of presenting different varying hardness levels by using different belt-winding speeds was effective to identify the hardness of the five kinds of objects.

Stimulation	Rate of the subject's answer [%]		
	Hard	Equal	Soft
S_1	4	12	**84**
S_2	16	16	**68**
S_3	24	**46**	30
S_4	**78**	22	0
S_5	**88**	12	0

Table 4. Identification results of hardness.

4. Feedback device of temperature sense (TFB device)

4.1. Temperature sense

A combination of cold and warm sensations is called a temperature sense. The temperature sense differs at each body site even when an object of the same temperature is touched. In addition, when the skin is exposed to extreme heat or cold, the pain along with the risk of burns arises. Therefore, it is necessary to regulate the temperature of the TFB device.

4.2. Peltier element and thermocouple

The Peltier element is an electronic device that enables both cooling and heating by applying a voltage on the basis of the Peltier effect, and temperature is adjustable by regulating the voltage. The TFB device has a Peltier element; thus, the temperature sense is transferred to the user. The Peltier element used in the TFB device is "TEC1-12708" made by HB Electronic Components.

A thermocouple is a temperature sensor on the basis of the Seebeck effect. The thermocouple is attached to the silicone finger mounted on the fingertip of the myoelectric prosthetic hand to measure temperature of an object when the fingertip touches the object. A K-type thermocouple "AD-1214" made by T&D Corporation is used in this study.

4.3. Overview of TFB device

Figure 6 shows the developed TFB device and its attached state on the user's upper arm. The Peltier element is attached to the inside of an aluminum board, and in order to raise a heat dissipation efficiency, a radiation sheet and a heat sink are attached to the other side of the aluminum board. The TFB device is attached on the upper arm of the user in contact with the user's skin, and the temperature sense that is corresponding to the temperature detected by the temperature sensor at the fingertip is transferred in the upper arm of the prosthetic hand user.

The device is small enough with the dimensions 50 mm (width), 60 mm (depth), and 17 mm (height).

When the TFB device is used for a long period, the accumulated heat causes a high temperature of the TFB device. Therefore, the continuous operating time of the TFB device is limited to 5 s.

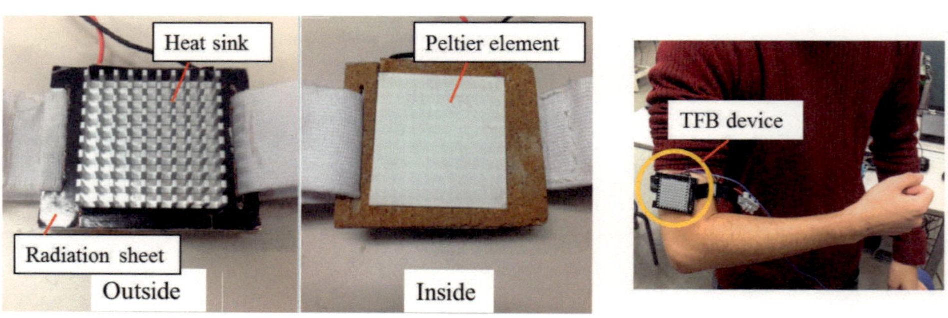

Figure 6. TFB device and its attached state.

When the TFB device decreases the temperature, it enables refrigeration of surface temperature to a minimum of 15°C for 5 s. On the contrary, when the TFB device increases the temperature, it enables heating of surface temperature exceeding 50°C. Hence, the temperature of the TFB device is limited to 40°C for safety.

4.4. Temperature prediction

After the myoelectric prosthetic hand contacts with the object, the temperature sensor needs a long time for measuring the temperature. Therefore, a temperature prediction is performed to shorten the measurement time.

Let relationship between a sensor output $y(t)$ and a temperature variation $u(t)$ be given by the following equation in transfer function expression:

$$Y(s) = G(s)U(s) \tag{7}$$

To determine the transfer function $G(s)$, the parameter identification was performed. Furthermore, let ΔT denote the predicted temperature variation and $\hat{y}(t)$ denote the estimated output, and then the estimated output is calculated as follows using the identified transfer function:

$$\hat{Y}(s) = \frac{0.89\,s^3 + 0.66\,s^2 + 0.49s + 0.06}{s^4 + 2.54\,s^3 + 1.7\,s^2 + 0.8s + 0.06}\Delta T \tag{8}$$

Let $\varepsilon(t)$ denote the error between the estimated output $\hat{y}(t)$ and sensor output $y(t)$, which is presented by $\varepsilon(t) = \hat{y}(t) - y(t)$. The predicted temperature variation ΔT is updated by using $\varepsilon(t)$ as follows in each sampling period:

$$\Delta T_{new} = \Delta T_{old} + K \cdot \varepsilon(t) \tag{9}$$

where ΔT_{old} is the predicted temperature variation one step before, ΔT_{new} denotes an updated predicted temperature variation, and K is an arbitrary constant, which was chosen as $K = 70$.

Let T_0 denote the room temperature, and then predicted temperature T is given as follows:

$$T = \Delta T_{new} + T_0 \tag{10}$$

The predicted temperature is updated to reduce the error when the sensor detected a temperature variation; thus, the temperature when reached to the equilibrium state is given by Eq. (10). As a result, the sensor can detect the temperature of the object in a short time.

The verification experiment was performed to verify an effectiveness of the proposed temperature prediction algorithm. In the experiment, under the room temperature of 26°C, the sensor touched an object of 40°C. Then, the temperature was predicted by the proposed algorithm. **Figure 7** shows the result.

The result showed that prediction of the temperature in a short time is possible by using the proposed temperature prediction method.

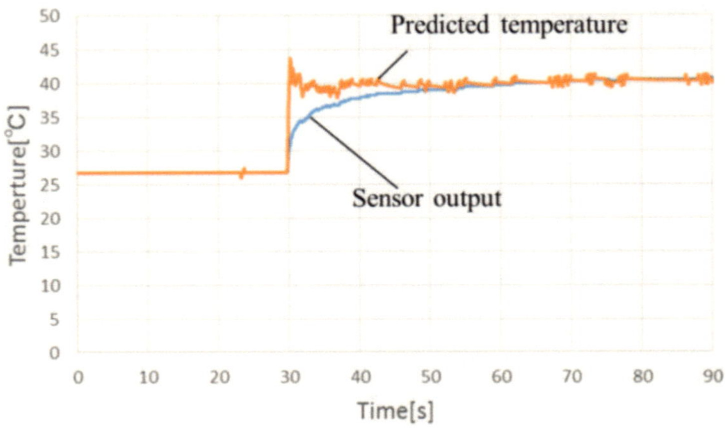

Figure 7. Result of the temperature prediction for the object of 40°C.

4.5. Temperature sense investigation

The temperature sense of an individual differs at each body site. For example, the temperature senses when an object with the same temperature is touched with a finger and with the upper arm are different. Therefore, in order to investigate the difference in the temperature sense between the fingertip and the upper arm, a temperature sense investigation was performed. Thus, the temperature experienced by the upper arm which is equivalent to the temperature experienced by the fingertip was determined as feedback temperature to the upper arm.

The temperature which should be presented at the upper arm and the voltage which should be applied to the TFB device were determined in [11] as follows. The relationship between the temperatures experienced by the fingertip and by the upper arm was interpolated. Then, the relationship between the predicted temperature at the fingertip T and the target temperature in the upper arm T_t and the relationship between the predicted temperature at the fingertip T and the input voltage to the TFB device V were approximated as shown in Eqs. (11) and (12), respectively, by using the least squares method:

$$T_t = 0.00008\,T^4 - 0.006\,T^3 + 0.1841\,T^2 - 1.1439T + 16.902 \tag{11}$$

$$\begin{aligned} V = {} & 1.578 \times 10^{-7}\,T^6 - 2.135 \times 10^{-5}\,T^5 + 1.122 \times 10^{-3}\,T^4 \\ & - 2.894 \times 10^{-2}\,T^3 + 0.3819 \times T^2 - 2.543T + 8.743 \end{aligned} \tag{12}$$

4.6. Closed-loop control system

4.6.1. Construction of the control system

In the previous section, the developed TFB device was controlled in an open-loop system, in which the constant voltage computed from Eq. (12) was applied to the TFB device. In the case where the voltage is continuously provided to the TFB device, the temperature keeps increasing or decreasing. Hence, it is impossible to maintain the temperature using this control system. Therefore, the continuous operating time of the TFB device was limited to 5 s.

To use the TFB device continuously, a closed-loop control system was constructed. For this purpose, another temperature sensor was additionally attached on the surface of the Peltier element of the TFB device. The target temperature in the upper arm T_t corresponding to the predicted temperature at the fingertip of the myoelectric prosthetic hand T is determined by Eq. (11). Then, the input voltage to the TFB device is determined by a PID controller. The PID gains were determined by trial and error and chosen as $K_P = 1$, $K_I = 0.05$ and $K_D = 0.01$.

4.6.2. Operation check of TFB device

In order to verify the effectiveness of the closed-loop control system with PID controller, experiment was performed. In the experiment, the target temperature is suddenly decreased from 40 to 15°C. The transition of the temperature of the TFB device and the input voltage to the TFB device are shown in **Figure 8**.

As shown in **Figure 8**, the results showed that the constructed closed-loop system enabled the adjustment of the temperature of the TFB device according to the temperature change and also enabled a long-time continuous operation of the TFB device.

4.7. Temperature identification experiment

In order to verify the performance of the TFB device controlled in the closed-loop control system, a temperature identification experiment was performed. The usefulness of the TFB device controlled by the closed-loop system is objectively verified with a psychophysics experiment method.

4.7.1. Overview of the experiment

In this experiment, the myoelectric prosthetic hand was not used, but each temperature was input directly to a computer, and then the temperature identification experiment using the TFB device was conducted.

The temperature of 30°C was used as standard stimulation, and five kinds of temperature, 28, 29, 30, 31, and 32°C, were used as comparative stimulation. The following describes the experimental procedure:

1. TFB device is attached on the upper arm of the subject.
2. An experimenter inputs the standard stimulation (30°C) to the computer, and standard stimulation is presented to the subject by the TFB device.
3. An experimenter inputs the comparative stimulation that is randomly selected from the five kinds of temperatures to the computer, and comparative stimulation is presented to the subject by the TFB device.
4. The subject answers which temperature is higher or whether the two are almost same.
5. 25 sets of the operations [(2)–(4)] are performed.
6. The operations [(2) and (3)] were replaced, and 25 sets of the operations [(2)–(4)] are performed.

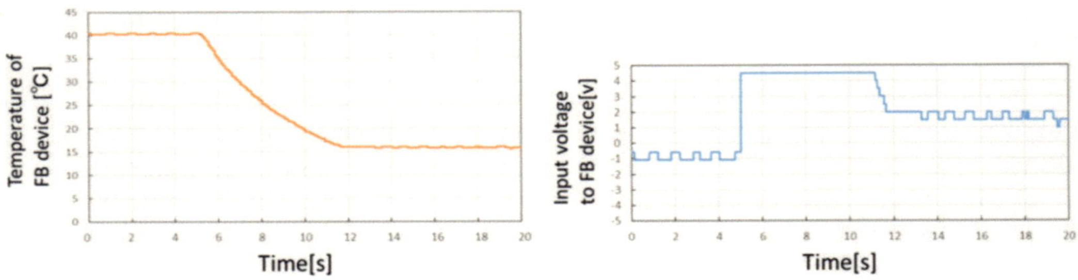

Figure 8. Temperature of the TFB device and input voltage to the TFB device.

Temperature of comparison stimulus [°C]	Rate of the subject's answer [%]		
	Hot	Equal	Cold
28	0	14	**86**
29	4	36	**60**
30	6	**82**	12
31	**60**	34	6
32	**86**	10	4

Table 5. Identification results of temperature.

In the operations, all the comparative stimulations were used 10 times in random orders. The experiments were performed for five healthy subjects in their 20s.

4.7.2. Results

Table 5 shows the results of experiments. The bold numbers show the ratio of correct identification of the stimulations.

From the results on the temperature identification experiment, one can see that the proposed method of presenting temperature by the TFB device controlled in the closed-loop control system was effective to identify the five kinds of temperatures.

5. Integration of the sensory feedback devices

5.1. New myoelectric prosthetic hand

To improve the operability of the myoelectric prosthetic hand, a new myoelectric prosthetic hand was designed and built by imitating the commercial prosthetic hand, which is shown in **Figure 9**. The pressure sensor and the temperature sensor were attached to the fingertip of the thumb and index finger of the prosthetic hand, respectively. Thus, this myoelectric prosthetic hand makes it possible to detect the force and temperature when the prosthetic hand holds an object.

5.2. Two-sensory feedback device

Finally, the FFB device and the TFB device were united, and a two-sensory feedback device was built, which is shown in **Figure 10**. The dimensions of the device are 75 mm (width), 82 mm (depth), and 34 mm (height).

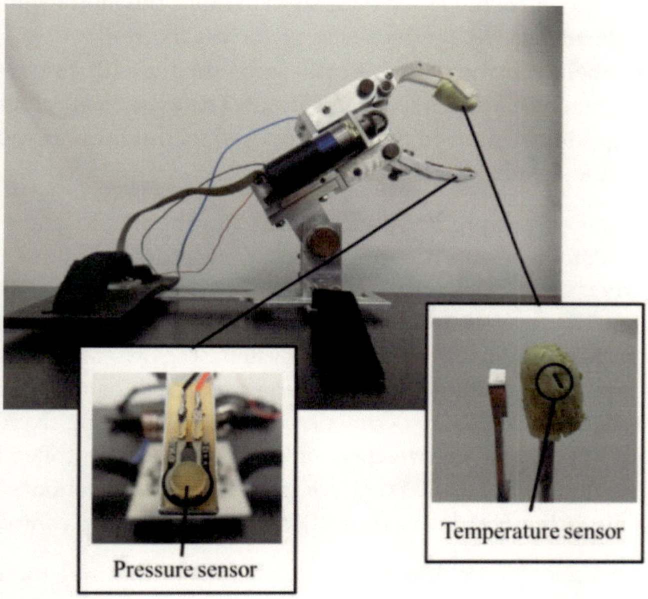

Figure 9. New myoelectric prosthetic hand.

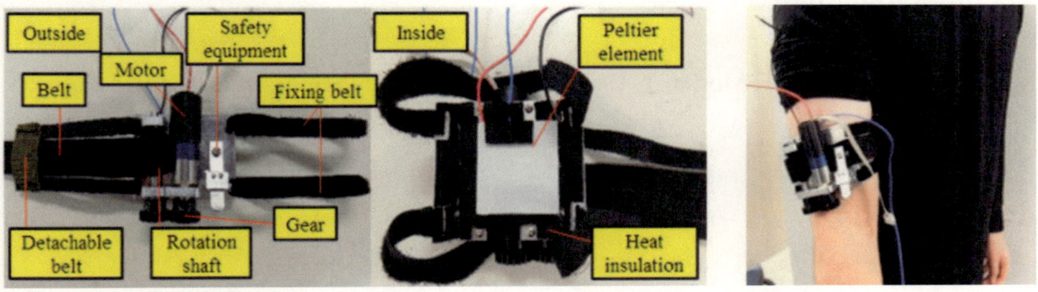

Figure 10. Two-sensory feedback device and its attached state.

6. Conclusion

In this study, force feedback device (FFB device) and temperature feedback device (TFB device) were proposed and built. When a user of a myoelectric prosthetic hand grasps an object, the FFB device provides pressure to the user's upper arm by winding a belt using a motor, and the TFB device presents the temperature sense to the user's upper arm using the Peltier element.

In the FFB device, the hardness of the object was estimated by a pressure sensor attached on the fingertip of the myoelectric prosthetic hand, and a reference input was produced by a reference input creation model according to the hardness. In addition, a self-tuning PID controller was employed to control the FFB device so as to make the motor's output angle follow the reference input. Furthermore, the hardness of the grasped object was presented by the winding speed of the belt. Hardness identification experiment to distinguish among the five kinds of springs of different hardness was carried out. The experimental results on the hardness identification experiment showed that the proposed method was effective to identify the hardness of the five kinds of objects.

In the TFB device, a temperature prediction algorithm was proposed for short-time temperature detection. Then, based on the results of the temperature sense investigation, the corresponding temperature sense when the object was touched by a fingertip was transferred to the user's upper arm by the TFB device. However, it was difficult to operate the TFB device continuously because this device was controlled in an open-loop control system. To solve this problem, a closed-loop control system was constructed for the TFB device and was tested for sudden change of the temperature. Temperature identification experiment to distinguish among five different temperatures was carried out to verify the effectiveness of the TFB device controlled in the closed-loop control system. The experimental results on the temperature identification experiment showed the sufficient capability of the TFB device controlled in the closed-loop control system.

In addition, a new myoelectric prosthetic hand was built to improve the operability of the myoelectric prosthetic hand. Finally, two-sensory feedback devices were united, and a two-sensory feedback device was built.

Acknowledgements

The author thanks Mr. T. Morita, Y. Ueda, and M. Isobe for their assistance in experimental works.

Author details

Chiharu Ishii

Address all correspondence to: c-ishii@hosei.ac.jp

Department of Mechanical Engineering, Faculty of Science and Engineering, Hosei University, Japan

References

[1] Clippinger FW, Avery R, Titus BR. A sensory feedback system for an upper-limb amputation prosthesis. Bulletin of Prosthetics Research. 1974;**Fall**:247-258

[2] Antfolk C, D'Alonzo M, Rosén B, Lundborg G, Sebelius F, Cipriani C. Sensory feedback in upper limb prosthetics. Expert Review of Medical Devices. 2013;**10**(1):45-54

[3] Mohamad Hanif NHH, Chappell PH, Cranny A, White NM. Vibratory feedback for artificial hands. In: Proceedings of International Conference on Electronics Computer and Computation; 2013. pp. 247-250

[4] D'Alonzo M, Dosen S, Cipriani C, Farina D. HyVE: Hybrid vibro-electrotactile stimulation for sensory feedback and substitution in rehabilitation. IEEE Transactions on Neural Systems and Rehabilitation Engineering. 2014;**22**(2):290-301

[5] Otsuka A, Tsuji T, Oki S, Sakawa M. Development of a feed-back-system for thermal sensation. Bulletin of the Japanese Society of Prosthetic and Orthotic Education, Research and Development. 2001;**17**(2):135-138 (in Japanese)

[6] Morimitsu H, Katsura S. Heat inflow control of Peltier device based on heat inflow observer. In: Proceedings of SICE Annual Conference; 2010. pp. 996-1001

[7] Yamamoto T, Omatu S, Ishihara H. A construction of self-tuning PID control system. Transactions of the Society of Instrument and Control Engineers. 1989;**25**(10):1069-1075 (in Japanese)

[8] Harada A, Nakakuki T, Hikita M, Ishii C. Robot finger design for myoelectric prosthetic hand and recognition of finger motions via surface EMG. In: Proceedings of the 2010 IEEE International Conference on Automation and Logistics; 2010. pp. 273-278

[9] Kikuchi T, Ishii C. Identification of finger operation using support vector machine and control of myoelectric prosthetic hand based on integrated electromyogram. In: Proceedings of the 2014 IEEE International Conference on Robotics and Biomimetics; 2014. pp. 1272-1277

[10] Morita T, Ishii C. Development of sensory feedback device for myoelectric prosthetic hand to provide hardness of objects to users. Journal of Robotics and Mechatronics. 2016;**28**(3):361-370

[11] Ueda Y, Ishii C. Development of a feedback device of temperature sensation for a myoelectric prosthetic hand by using Peltier element. In: Proceedings of the 2016 International Conference on Advanced Mechatronic Systems; 2016. pp. 488-493

Chapter 4

Optimizing User Integration for Individualized Rehabilitation

Raviraj Nataraj

Additional information is available at the end of the chapter

http://dx.doi.org/10.5772/intechopen.70267

Abstract

User integration with assistive devices or rehabilitation protocols to improve movement function is a key principle to consider for developers to truly optimize performance gains. Better integration may entail customizing operation of devices and training programs according to several user characteristics during execution of functional tasks. These characteristics may be physical dimensions, residual capabilities, restored sensory feedback, cognitive perception, or stereotypical actions.

Keywords: prostheses, exoskeletons, rehabilitation

1. Introduction: User 'feel' for rehabilitation devices

#FeelBetterDoBetter was a recent campaign slogan by apparel giant Nike, suggesting how sportswear can provide a 'feel' that leads to better athletic performance. Does such a concept apply to prosthetic devices and improved rehabilitation? Is feel a matter of user perception, product customization, or literally the sensation of touch? Great technological strides have been made to develop assistive devices and paradigms that restore motor capabilities of individuals with physical impairment. Augmenting function with prosthetic limbs after amputation [1] and neuroprostheses following spinal cord injury [2] are clear examples where the biological-technological interface has notably advanced. Smart-prosthetics are continually being equipped with greater computational processing and actuation methods to accurately control movements involving several degrees of freedom [3]. Electromotor prosthetic limbs can now sufficiently power desired motor actions while maintaining compact form for increasingly biomimetic implementation [4]. Neuroprostheses for motor restoration now

includes greater access and finer control of paralyzed muscles activated by advanced stimulation electrodes [5, 6]. While further advancements in augmenting the capacity of movement for those with motor dysfunction is still sought, it is clear there has been notable advancement for assistive device users to 'do'.

How then is the full potential of 'do' realized with movement rehabilitation? The Nike slogan would imply it is a matter of better 'feel'. What defines **how** a user feels when using a rehabilitation device? A literal interpretation may be engineering novel sensations of touch or body movements to the user through the device interface itself. Such an example is a prosthetic hand following amputation that connects mechanical sensors to transduce physical sensations to neural activations and allow users to perceive touch or movement [7, 8]. With intact physiological systems interfacing with external objects, such sensory feedback is meant to facilitate the identification of object shape, size, weight, and texture [9]. This information should allow better manipulation of objects with appropriate motor planning, online adaptations, and final goal achievement [10]. Another interpretation of feel is how a user perceives their ability to use the device. This may initially be reflected in usability surveys [11]. This could imply ease-of-use or sufficient capabilities to do activities of daily living. For the former, the perception of ease-of-use can be highly subjective. For the latter, the user's perception of value could be in terms of what new tasks they cannot otherwise perform or at what new level they can perform certain tasks. Is perception of ability simply a matter of effort, or in the case of devices that are restoring functions in a biomimetic way, is it truly a matter of user integration?

Integration can also be described on multiple levels. In the case of a device, the user may see the device truly as an extension of oneself. Certainly, the ultimate objective for integration would be to provide means that maximize biomimicry, including allowing the user to feel embodiment to the device [12–14]. Not only should the device be anthropomorphic but it should also provide to the user sensory feedback, including touch and kinesthesia [15]. Finally, the device should reflect user intent. In cognitive neuroscience, this would entail having a sense of 'agency', or belief one is the true author of one's movements [16]. Understanding how better to integrate the user with a device may lead to better performance of a specific movement function. Therefore, it may be critical to identify the design specifications of a rehabilitation device that improves user integration to better execute the desired movement. With movement execution, feedback describing system kinematic states [17, 18] is fundamental at a device level. But allowing users to 'feel' more integrated with operation may be the key better performance by user-driven devices.

Areas of advancement to improve device 'feel' include: (1) using novel modes of feedback, (2) computational intelligence for processing control and identification, (3) targeted user-based command strategies to initiate and modulate control of actions, and (4) identifying optimal movements to be executed. For novel modes of feedback, not only could new sensory modalities be biophysically restored, but device controller and rehabilitation protocols may better leverage how users employ intact sensory feedback capabilities. Users may rely on intact visual, audio, and haptic feedback to negotiate movements and modulate their control of a device or actions during a rehabilitation training protocol [19]. New control structures are taking advantage of increasingly powerful computational capabilities and soft computing approaches [20]. These approaches may require less accurate descriptions of system identification and can readily update parameters to adapt online in the presence of new performance data. Driving operation of assistive devices according to user commands for better performance has typically involved selection of desirable user cues, tuning operational sensitivity to those cues, and facilitating heuristic learning [21]. Simple cues may allow users to invoke switch-type control to initiate operation with a mechanical trigger like a button push [22]. Extension to biomimetic operation would rely on more graded control and bioelectric signaling from the user such as electromyographic (EMG) or electroencephalographic (EEG) recordings [23, 24]. Subsequent refinement and tuning between developers and users for customization can be largely expected to generate satisfactory empirical results. Thus, customizing devices and protocols to each user is an important design consideration to produces the best movement results for the user. Certainly, considering physical dimensions to better fit a device to someone is required for basic assimilation. However, the residual capabilities of the user may not be as well considered except to define a functional baseline from which the user is expected to improve with practice. This approach leaves the onus largely on the user to learn how to better operate the device or diligently engage in a protocol.

In this chapter, we will suggest rehabilitation approaches with examples that consider 'feel' in development of devices and protocols that better integrates users for improved performance. In Section 2, we outline the major components in a user-driven rehabilitation device or protocol and their bases of interaction. In Sections 3–5, we discuss alternative methods and considerations to better incorporate the user into assistive devices or rehabilitation protocols. We first postulate an example of controlling a gait-restoration exoskeletons that could optimally relate volitional actions of the user with computation of device control parameters that optimize performance. We then discuss methods that consider how user perception of agency over a device may affect performance. The discussion context will be a proposed experiment to optimize an individual's ability to perform grasping actions with a virtual hand prosthesis based on cognitive agency. Finally, we suggest a rehabilitation paradigm to minimize injury when they consider individualized physiological responses to movement activity.

2. System view of user-centered rehabilitation

There are three major components to consider for a system that restores or rehabilitates movement function (**Figure 1**, top). First, there is the biomechanical plant which is composed of the physical body of the person executing a movement task objective within an environment, including objects to be manipulated. In this formulation, we interpret the fundamental movement goal as a fixed property of the system plant. The physical body includes the size and shape characteristics of the limbs to be moved, perhaps the entire body as with ambulation. Second, a device may assist in executing the movement or provide constraints and cues to the user on how better to perform the function. Third, the user volitionally commands themselves and/or the device to drive movement operation of the entire system. The extent of the drive depends on residual capabilities of the user, contingent on the disorder, the capabilities of

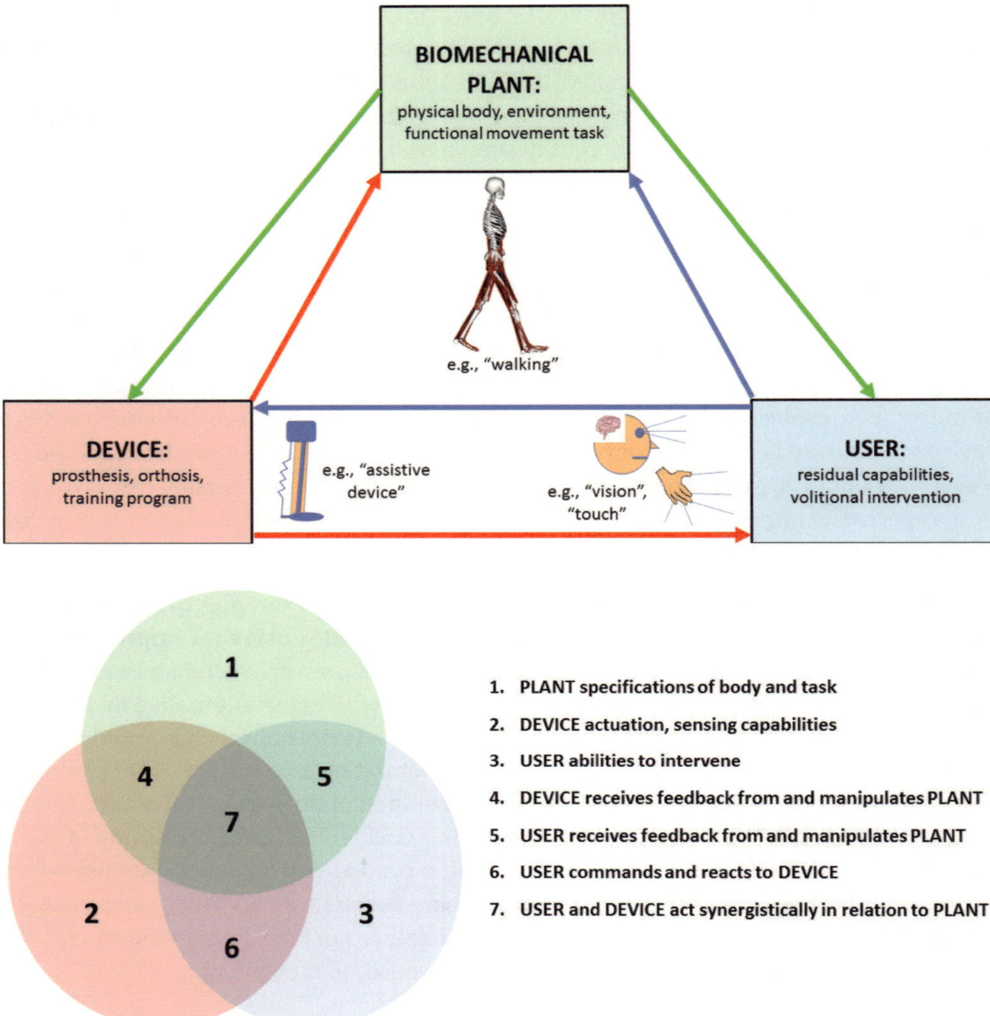

Figure 1. Top: Three major components for control system that rehabilitates movement function. Bottom: Component areas of overlap having distinct research thrusts.

the device, and the user-device interface. For example, an individual with stroke performing walking may largely be servicing the movement with their own residual capabilities. They may utilize an assistive device such as a spring-loaded ankle-foot orthosis (AFO) for propulsive boost during gait push-off. The user may further govern execution of gait with their vision and the haptic feedback they receive at their feet or even an additional assistive device such as a walking cane.

All three major system components (plant, device, user) must act in concert to successfully maintain or execute movement. The synergistic interactions between each component represent areas of research investigation to be developed to better restore movement performance.

In **Figure 1** (bottom), areas 1, 2, 3 are specifications for each fundamental component previously described. Area 4 signifies how the device utilizes its onboard sensing capabilities to receive feedback about actions of the plant, and in turn utilizes its actuation capabilities to manipulate the plant. Depending on the level of assistance the device provides, it likely requires proportionate information about the states of the plant. For movement, these feedback states are likely kinematics. The device may utilize potentiometers to measure joint angles or inertial sensing units that integrate accelerometers, gyroscopes, and magnetometers to derive real-time estimates of global orientation [25]. Load sensors may be utilized to discern pressure or force information between plant and environment such as ground-foot contact [26] or hand grasp loads onto objects [27]. Device actuation may be more passive as with AFO assistance from spring-loads being applied proportional to displacements ala Hooke's law ($F = -kx$) and mediated through well-executed clutching [28]. Device actuation may be more active such as functional neuromuscular stimulation (FNS) to activated paralyzed musculature and restore standing balance capabilities [29] or electromotor actuation to restore walking [30] following spinal cord injury (SCI).

Area 5 similarly denotes how the user may rely on his or her own residual sensory capabilities (such as vision or touch) to interpret states of the plant and can enact residual motor capabilities to further manipulate the plant independently or with the device. For the purposes of rehabilitation, user-volitional capabilities alone are not sufficient to achieve the desired levels of function. Thus, a device or training program to enhance overall function is necessary to desirably improve movement performance. The relative user versus device contribution function can vary considerably depending on application. The user typically drives most of the action, which is only enhanced by the device, for gait following stroke. Alternatively, it may be the device that makes movement even possible such as with FNS or powered exoskeletons to restore standing and walking following complete SCI.

Area 6 is the fundamental link between user and device. Often, the device operates in a preprogrammed manner according to distinct phases of the movement (e.g., gait cycle), or the user will command operation of the device (e.g., button push, tilting action, signal threshold exceeded). In the latter case, the onus is on the user to monitor the plant and select when the device should intervene. Ultimately, the user responds not only to his or her own observations of the plant but must react to the actions of the device as well. The device may also provide information to the user about the plant that the user does not otherwise have. In this case, the device may restore sensory feedback (touch or movement) to the user about body-environment-task interactions [31, 32]. The device may do so through neural activation to restore basic sensory capabilities such as vision, hearing, touch, or kinesthesia [33]. Or the device may provide supplementary sensory cues to alert the user to events or cautionary indicators related to the movement. For example, vibrational cues have been used for training individuals on how better to use a device [34], and audio cues have been provided for assisting individuals in balance function [35].

Moving forward, it is research initiatives in Area 7, that designs simultaneous synergies between plant, user, and device that may truly optimize rehabilitation for better functional performance. Considering how best to integrate the user may be most key to this end. Mechanical descriptions of the plant and target movement is necessarily done for designing the functional specifications of an assistive device. But integrating the user is comparatively more challenging without *a priori* indication of their responses (mechanical, physiological) using a device. It is often expected that the user will adapt and train with the device, and device parameters can be tuned by developers according to empirical patterns, ad hoc adjustments, and anecdotal feedback from the user. These approaches are understandable and imminently practical, but are they ultimately optimal? Rehabilitation approaches that better consider user integration in the early stages of development may better produce the synergy between plant, user, and device that maximizes performance. In the next three sections, we postulate examples of rehabilitation approaches to maximize movement performance by considering: device operation according to user actions, optimizing user sense of agency over a device, and identifying optimal movements from user-specific physiological responses.

3. Enacting device control from user actions

A biomechanical rehabilitation system with great potential for better incorporating user actions into enacted device operation is gait-restoration by powered exoskeletons. To overcome clinical barriers to translation, powered lower-limb orthoses for gait following SCI should be light-weight, cosmetic, and well-integrated with standard wheelchairs. With these design constraints, however, joint degrees-of-freedom and torque magnitudes are typically limited. Users must then utilize walking aids such as crutches or canes for balance and support. While exoskeleton-assisted gait is achievable, the observed motions are staggered and expend energy inefficiently for the user and orthosis. Feedback control schemes may be better designed to synergistically combine functional capabilities and actions of the user and worn exoskeleton to produce gait patterns that are more natural and energy cost effective.

An *optimal* control structure would consider full body dynamics in generating exoskeleton motion patterns that resemble able-bodied gait while also operating at minimal energy cost [36]. The motion patterns may be limited according to constraints on actuation by both the user and the device. These actuation constraints may be in terms of how many degrees of freedom are actively under control and the maximum torques that can be generated. It is, in part, due to the constraints that users must support and balance the actions of the device with upper-body walking aids such as crutches. The user typically must reactively respond to the actions of the device after the user initially triggers the device on a step-by-step basis. Upper-body actions have previously been used for command triggering the next step of exoskeleton operation [22]. Ideally, controllers would be online modulate operation of the exoskeleton according to proactive user actions of the support aids to generate the most efficient (optimal) walking patterns. An optimal feedback controller could continuously vary feedback gains over the walking cycle to synergistically combine actions of the user upper-body and lower-limb orthosis. This could provide better performance and a more natural command interface. This user-centered objective in controller development would be to utilize user-commanded crutch motions in real-time to efficiently regulate exoskeleton actuation. In **Figure 2**, we show a block-diagram schematic of how such an exoskeleton control system may be considered.

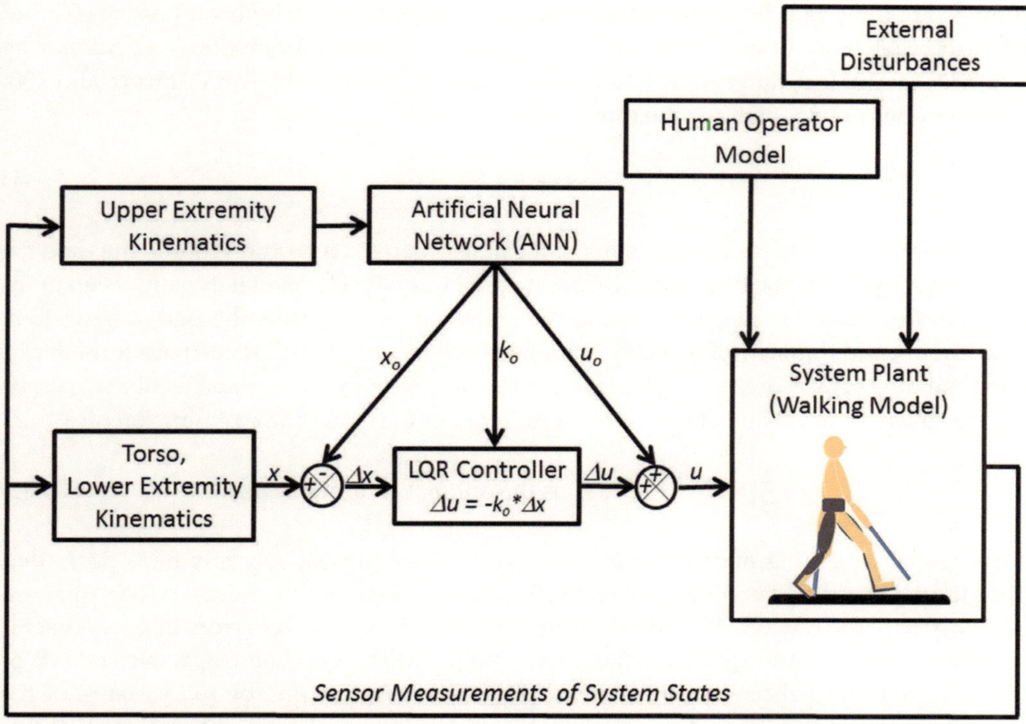

Figure 2. Example of controller system to modulate assistive device operation according to user actions and generate optimal walking function.

The linear quadratic regulator (LQR) is a classical optimal control approach to regulate the dynamics of a linear system. The LQR can also be utilized on non-linear dynamical systems such as bipedal human gait by approximating as a linear time-varying system. Extending such an approach could provide a foundation for optimal control of powered gait devices such as exoskeletons. Exoskeletons to restore or augment gait function have previously employed time-varying proportional-derivative (PD) control [30]. However, gains and setpoints for closed-loop joint control were empirically fitted to human angle-moment data and consequently not unique nor optimal toward specific performance objectives. This is unlike an optimal control approach where parameters are uniquely identified for minimizing a specified cost function such as controller effort while better tracking desired motion trajectories.

Bipedal walking can largely be simulated with a sagittal plane model. A minimal representation may consist of seven rigid bodies, connected by six bilateral joints of the hips, knees, and ankles [37]. For full-state description, there would be nine kinematic generalized coordinates: global anterior-posterior and superior-inferior position coordinates of the hip joint, torso orientation, right/left hip angles, right/left knee angles, and right/left ankle angles. In this model, contact between feet and ground can be modeled with spring-damper elements uniformly distributed along each foot sole [37] and include some passive joint moments to provide realistic joint motions and limits. Reasonable passive moments include basic stiffness (1 N-m/rad) and damping (1 N-m-s/rad) through normal ranges of motion, and then high stiffness (1000 N-m/rad) near range of movement limits. Modes to generate equations of motion (e.g., Autolev by Online Dynamics Inc., Sunnyvale, CA) can be readily integrated with MEX functionality [38] to compute system dynamics in the form:

$$\dot{x} = f(x, u) \tag{1}$$

Where x is the full set of states, consisting of the nine generalized position coordinates and the corresponding nine velocities, and u are the six joint torques. The model dynamics equations should be formulated as twice differentiable to facilitate use of gradient-based optimization, linearization, and implicit solvers [39]. For a linear system in general state-space form in discrete time: $x_{i+1} = A_i x_i + B_i u_i$, an LQR formulation would provide an optimal feedback gain (k) solution given a control law of $u_i = -k_i x_i$ that minimizes the quadratic cost function of:

$$J = \sum_{k=0}^{\infty} \left(x_k^T Q x_k + u_k^T R u_k \right) \text{as the optimal control problem.} \tag{2}$$

The diagonal weighting matrices Q and R quantify how to relatively minimize deviations from desired states and controls, respectively. The matrices serve as control design parameters to specify the relative desired effects for better tracking (smaller errors in states) versus less effort (lower joint torques). Each diagonal term corresponds to a respective state (i) or control (j) that normalizes its contribution to the objective function by the variance of the respective optimal trajectory. For variance, the matrix diagonal terms are set equal to the standard deviation (σ) values of the optimal trajectory for each state or control. These σ values allow us to perform quadratic normalization of each state and control term for units and magnitudes of variation during gait as follows:

$$Q_{ii} = \varphi \left(\frac{1}{\sigma_{x_i}} \right)^2 \text{ where } i = 1 \text{ to } 18 \text{ states} \tag{3}$$

$$R_{jj} = \left(\frac{1}{\sigma_{u_j}} \right)^2 \text{ where } j = 1 \text{ to } 6 \text{ joint} - \text{torque controls} \tag{4}$$

To uniformly vary weighting toward tracking over effort, only the Q_{ii} terms need be subsequently and multiplied by a single Q/R ratio, φ. This ratio serves as the single controller design parameter. It is assumed with $\varphi = 1$, there is equal consideration of tracking and effort in controller performance. The optimal feedback gain matrix, k, can then be solved across time using a discrete, periodic Riccati equation solver [40].

For "exoskeleton walking," which typically requires volitional upper-body support, feedback states would need to include kinematics of the torso and upper-extremities as well. The upper-extremities could include extensions of the crutches that make ground contact, converting the model to a type of quadruped. Alternatively, if arm support actions were more suspensory, as with a walker, user 'control' could be equivalently modeled as loads at the shoulders [41, 42]. However, the upper-extremities with crutch walking would allow controller developers to utilize volitional motion of the upper extremities of the user to signal gait intentions. From these gross kinematic actions, desirable operation of the powered exoskeleton may be better identified. An LQR controller could be defined for the entire plant (include upper and lower extremities). A family of LQR controllers for different sets of optimal whole-body kinematic trajectories could be generated offline. Each set could represent a different gait speed or various couplings between upper and lower body motions. In any case, the sets of trajectories should comprehensively cover the spectrum of potential motion behaviors for a given user. But to parse out upper extremity actions to trigger lower extremity control, two distinct control loops would be necessary. First, if the user (human operator model) has the freedom to modulate gait from their upper extremities, then one loop needs to map the expected or desired lower extremity control actions. These actions would be necessarily coupled from the observed/measured upper extremity feedback states. Some mapping structure (e.g., artificial neural network) is necessary to identify from the upper extremity kinematics what is the corresponding desired kinematic states (x_o), gains (k_o), and controls (u_o) for the exoskeleton. Beyond controller actions, the remaining actuating elements on the plant are the user (i.e., human operator) and external disturbances. Such a block diagram system for user-driven exoskeleton walking could be simulated if stereotypical actions for the human operator and environmental (external) disturbances could be reasonably modeled.

4. Optimizing user agency over device

User-driven devices are often judged by measured ability to restore or add functional capabilities following neuromuscular disorder or injury. Limb loss has a profound impact on an individual's capability to perform activities of daily living. Amputations of the upper-limb can be especially debilitating to one's ability to interact with their environment and feel physically engaged to the world. Prostheses for limb loss are first-level evaluated by simple physical fit and comfort [43]. But further criteria for artificial limbs is feeling natural and operating as real ones based on embodiment [44] or brain-level control [45]. Applied neurosciences are also creating prostheses to provide a better sense of interaction with the external world and to seamlessly take actions according to user intent. Neuro-applications have been developed to elicit motor and sensory signals at a peripheral neurophysiological level for novel electromyograms and restoration of touch [1, 8]. But more complete user integration with the device may reside at a cognitive level. The user should have a better 'feel' for the device, not just by literal touch restoration, but a sense they are controlling the device intuitively. These devices should be operated with sense of agency, or the perception the user is the true author of the device actions [14, 16]. In this section, we propose a framework to identify processes that optimize sensory-feedback hand prostheses according to agency. Such an approach may provide insight into design formulations that better integrate users with their devices, facilitate clinical acceptance, and ultimately produce skilful performance control that is more intuitive to the user.

For a user to inherently recognize the prosthesis as an extension of self, device operation should facilitate a sensorimotor basis of self-awareness. A key aspect of self-awareness is recognition of being the author of one's own voluntary actions and the related consequences.

This phenomenon is known as a sense of agency. It has been shown that cognitive processes underlying agency may be abstracted even in the presence of intermediaries such as prostheses [14]. Thus, a sense of agency may not require direct physiological embodiment, or body ownership, and can exist across novel instrumental associations. Both agency and body ownership can be present with a surrogate prosthesis, as shown with the rubber hand illusion [12, 44]. But the sense of agency may be reduced by altered embodiment. However, distortions in the causal chain that produce end-effector actions that are incongruent with user intention more notably reduces agency [14]. There is evidence that with increased agency there is greater ability to perform functional tasks [46].

There are innovative approaches restore sensory and motor pathways for better control of functional hand prostheses after upper-limb amputation. One such approach is targeted muscle reinnervation (TMR) where nerves that once controlled the lost hand are surgically reassigned to target sites upon denervated pectoral muscles [1]. Resultant EMG recordings from these sites represent motor commands to the missing hand that can be reliably utilized to drive a motorized prosthesis. And tactile manipulation of skin at these sites stimulates afferent pathways that elicit sensory feedback of the missing hand to the amputee. The biological neural-machine-interface created by TMR is a unique platform for investigating control of multi-sensory prosthesis [7, 31, 32] and potential cognitive frameworks of agency. Specifically, agency considerations may enhance the EMG motor command interface and augment how sensory feedback is perceived by the user. One could utilize a computer-generated virtual hand [47] to represent multi-joint prosthesis dynamics to be driven by descending motor commands recorded at TMR muscle sites. Tactile pressure and vibratory stimulation of reinnervated skin areas can activate feedback pathways that elicit illusions of touch [48] and kinesthesia [33] to be perceived as superimposed upon the computer hand. Procedures for perceptual mapping to restore robotic touch [12] can be used to identify locations and parameters for pressure and vibratory stimulation. The virtual hand could perform functional grasp tasks such as matching dynamic grip-load profiles [49, 50] with concurrent visual feedback of the hand. The hand could be driven by EMG command with concurrent sensory illusions provided according to hand motion and contact with virtual objects. The end of the grip loading profile will represent task completion, after which tones will be sounded at variable intervals to detect intentional binding as a measure of agency. Intentional binding is the perceptual compression of the time-interval between a movement event and sensory consequence with action is voluntary [51], i.e., greater agency. Ultimately, patterns of EMG and visual illusions for touch and movement can be preferentially selected based on agency. Efficacy of such an approach may reflect improvement in performance at various levels of prosthesis development. These levels include EMG classification accuracy of user motor intent, user ability to interpret sensory feedback, and user functional performance.

Myoelectric control has been a major focus area for improved hand prosthesis function to create user command interfaces that are intuitive and effective [21, 52, 53]. Notable work has demonstrated the advantages of EMG pattern recognition over amplitude-based command to provide intuitive prosthesis control [54]. While these methods are largely based on prosthesis operation, the motor command interface could be designed to co-maximize function and sense of agency. Typically, EMG data from target muscle sites are collected to identify specific grasp patterns based on feature extraction [55–57]. Several of these methods produce similar identification results provided an appropriate selection of features [58]. Subsequently, the real-time effectiveness of the pattern classifier is evaluated in testing environments that observe functional metrics such as task times and success rates [59] under proportional control [60]. We postulate that collection of EMG-driven prosthesis grasping data with an intentional binding paradigm [16] may facilitate improved classification and performance given user-preferred grasping patterns. We postulate the collection of data and implementation of a hand prosthesis based on agency in **Figure 3**.

It may be possible that preferential selection of electromyography data for pattern classification training from trials where the subject indicates greater agency will produce higher classification success rates for various functional grips (cylinder, flat, and precision) [61]. For amputee subjects, training data collection can be done as part of grip matching paradigm between intact hand and missing (imagined) hand [62]. Using standard classifier plus standard proportional control [63], users could undergo functional EMG-drive trials to execute open-to-close grips of the virtual hand according to visual cues for target speed and position. For each trial, the sensory illusion and virtual hand being visualized could be systematically altered. With each trial, the intentional binding metric with interval estimation would be evaluated to assess sense of agency. Optimal grip configurations for posture and movements may be identified that maximize agency and subsequently grasp performance. It is then reasonably hypothesized that subjects will more reliably generate EMG patterns for conditions in which they have greater agency. Ultimately, the EMG data used to train and test the pattern classifier, the grip kinesthetic illusions, and grip configuration illusions while having tactile feedback could all be preferentially selected according to user agency.

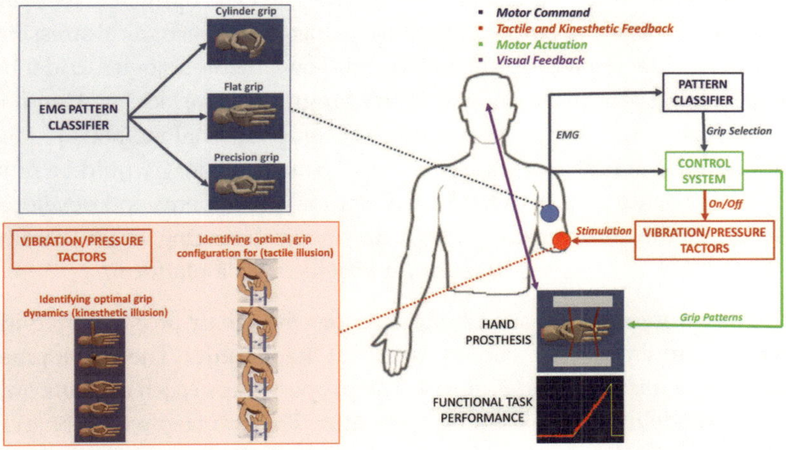

Figure 3. Left: Proposed data to collect for sensory-feedback prosthesis that considers optimization based on user's visual observations of preferred illusions. Right: Block diagram to evaluate functional performance of optimized sensory prosthesis.

5. Identifying optimal movements from individual physiology

Observable biomechanical characteristics such as kinematics, external forces, masses, lengths, inertias) can be well assessed for function, but the underlying physiological patterns may not be. Muscle force patterns are often estimated from optimization with the criterion of minimal muscle effort [64] following an inverse analysis [65] to determine joint moments. We hypothesize the value of customizing the minimization criterion based on observable physiological responses for an individual. In turn, a model-based approach could be developed for rehabilitation paradigms to identify movements that are optimal for an individual based on other criteria such as minimizing injury stresses. For the remainder of this section, we propose an example methodology to modify a musculoskeletal modeling environment based on person-specific physiological responses to identify optimal movements that potentially minimize shoulder injury for those using wheelchairs.

Disease or injury affecting mobility commonly requires afflicted individuals to excessively recruit the upper-body when ambulating with walking aids, executing seated transfers, or propelling a wheelchair. The repetitive and unnatural loading of the arms creates abnormal joint stresses that often precipitate shoulder pathologies. The spectrum of shoulder injury includes shoulder impingement syndrome, rotator cuff tears, glenohumeral instability, avascular necrosis, acromioclavicular joint degeneration, biceps tendonitis, and distal clavicle osteolysis [66]. The etiology of shoulder pathology for the movement disabled is typically associated with increased mechanical loading at the shoulder complex [67] and abnormal internal stresses at the shoulder joints [68]. Higher magnitudes of forces at the shoulder and poorly directed forces with inefficient motion can be detrimental to surrounding muscles and bones [69]. It has been suggested that proper technique may reduce progression of shoulder pathology despite high usage. Wheelchair athletes have fewer hospitalizations and no greater incidence of shoulder pathology [70], and focal strengthening of target muscles [71] and optimal maneuver techniques [72] can be vital in preventing shoulder injury. While increased mechanical stressing of muscles and bones of the shoulder are well correlated to pathology, methodologies to identify user-specific movement training patterns to mitigate injury progression are lacking. Methodologies that utilize person-specific physiological responses with computational methods to identify movements that are optimal in function and minimizing potential injury would be of high clinical value. This would provide platforms to identify optimal movements and develop customized training protocols for those movements. Integrating medical imaging, multi-scale musculoskeletal modeling, and advanced apparatuses could provide such a platform.

As an example, we describe such a methodology for wheelchair propulsion to identify optimal movements for minimizing functional stresses at the shoulder. The developmental flow of the proposed approach is shown in **Figure 4**. The proposed approach exploits unique advantages of imaging, modeling, and a training apparatus. The first step would be to create a custom testing apparatus with specialized instrumentation, computer-controlled actuation, and visualization facilities to simulate advanced propulsion functions. Feedback control of wheel resistance could serve to systematically simulate ridges, harsh surfaces, and inclines. This advanced biomechanical testing apparatus facilitates well-controlled investigation of wheelchair propulsion for consistently repetitive and functionally specific data for an individual.

http://dx.doi.org/10.5772/intechopen.70267

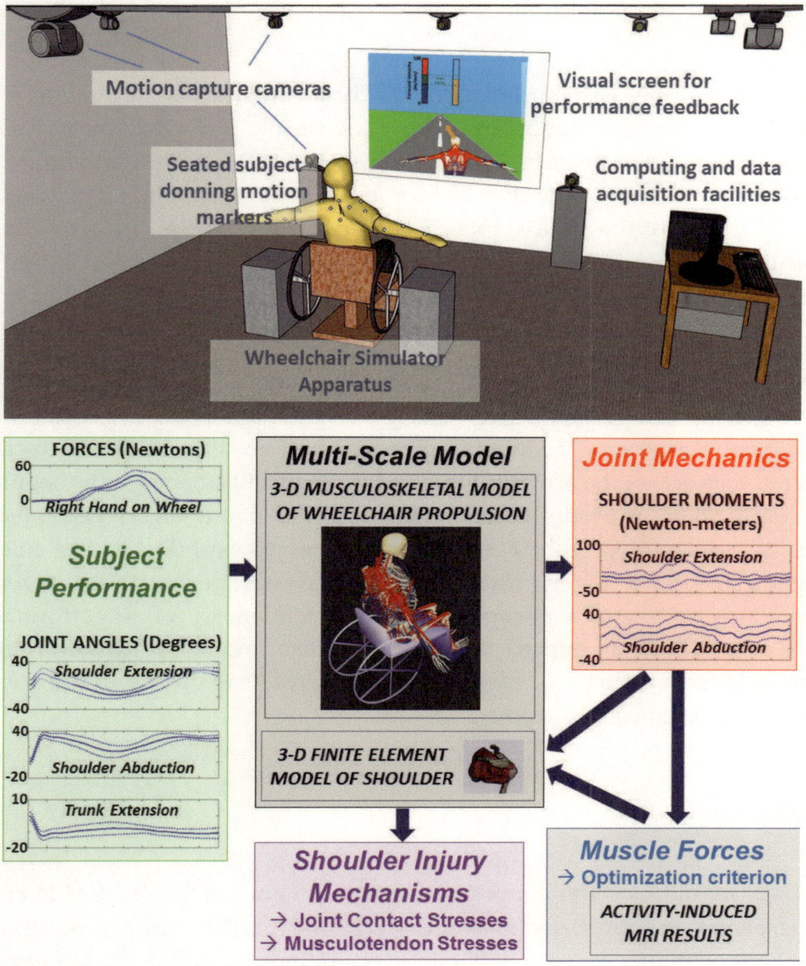

Figure 4. Top: Simulator for wheelchair propulsion with visual feedback and force and motion data capture. Bottom: Data flow to compute shoulder stresses by integrating high-resolution imaging and multi-scale musculoskeletal modeling.

The second step would be to identify injury stresses at the shoulder by integrating high-resolution imaging and multi-scale musculoskeletal modeling. Advanced medical imaging is typically used for physiological diagnosis of disease progression or structural characterization. Magnetic resonance imaging (MRI) can quantify activity-dependent muscle usage from contrast-detection of edema patterns following exercise [73, 74]. After a session of wheelchair propulsion, the motion and MRI data for that person could be input into a musculoskeletal model to compute shoulder stresses. The model would be multi-scale with global rigid-body dynamics used to compute muscle forces local finite element models (FEM) [75] to quantify local bone and muscle stresses that precipitate injury. The MRI data could be used to create outlines of the FEM models but also bias an objective function being minimized for optimality for the specific person. The optimization can have a multi-objective minimization function:

$$f = K_1 \sum_{i=1}^{\#muscles} b_i a_{i,model}^2 + K_2 \sum_{i=1}^{\#muscles} \sigma_{i,intrinsic}^2 + K_3 \sum_{i=1}^{\#bones} \sigma_{i,contact}^2 \quad (5)$$

The K-terms pre-multiply each summation based on experimenter discretion to relatively weight contribution of each criterion and ensure robust solution convergence. The first summation represents muscle activation levels at the shoulder. Minimization of this term desirably reduces muscle activity at the shoulder for locomotor *efficiency* [64]. The bias coefficients (b_i) weight *distribution* of the relative muscle edema to consider the strain results from MRI and to shift task requirements away from the shoulder muscles that the subject relatively overuses. The second and third summations are for explicit minimization of the injury mechanisms for intrinsic muscle and bone stresses, respectively.

Finally, a rehabilitation paradigm could be generated from this methodology to provide customized training to individuals for performing the optimal movements to minimize injury. The optimization routine should identify propulsion patterns that efficiently distribute functional requirements across the arms and torso while reducing stresses at the shoulders based on person-specific data. A proposed rehabilitation protocol may utilize visual feedback targets to train its users to modify their cyclic motor actions toward the biomechanical patterns that minimize injury. Such training results could provide new perspectives into sensorimotor adaptations that occur when balancing natural motor tendencies versus performance targets to minimize injury. Such an approach to identify sensorimotor principles of model-based learning for injury prevention is novel and readily extensible to other exercise science, sports performance, and motor control studies.

6. Conclusions

In this chapter, we outlined the importance in user integration for the next major advancement in the feel and performance of movement rehabilitation devices and protocols. We presented three case examples for user integration: (1) optimizing device operation according to user actions, (2) designing a prosthesis controller according to cognitive agency, (3) developing musculoskeletal modeling platforms based on person-specific physiology. There is great potential to further develop such ideas for new applications and in combination for improved rehabilitation. In any case, user-centered approaches for movement rehabilitation offers the greatest promise for user acceptance of such devices and programs beyond further technological breakthroughs for sensing, actuation, and biological interfacing.

Author details

Raviraj Nataraj

Address all correspondence to: rnataraj@stevens.edu

Department of Biomedical Engineering, Stevens Institute of Technology, Hoboken, New Jersey, USA

References

[1] Kuiken TA, Miller LA, Lipschutz RD, Lock BA, Stubblefield K, Marasco PD, Zhou P, Dumanian GA. Targeted reinnervation for enhanced prosthetic arm function in a woman with a proximal amputation: A case study. Lancet. 2007;**369**(9559):371-380

[2] Nataraj R, Audu ML, Triolo RJ. Center of mass acceleration feedback control of standing balance by functional neuromuscular stimulation against external postural perturbations. IEEE Transactions on Biomedical Engineering. 2013;**60**(1):10-19

[3] Miller LA, Lipschutz RD, Stubblefield KA, Lock BA, Huang H, Williams TW, Weir RF, Kuiken TA. Control of a six degree of freedom prosthetic arm after targeted muscle reinnervation surgery. Archives of Physical Medicine and Rehabilitation. 2008;**89**(11): 2057-2065

[4] Herr, H, Whiteley GP, Childress D. Cyborg Technology--Biomimetic Orthotic and Prosthetic Technology. 2003. Bellingham, Washington: SPIE Press

[5] Fisher LE, Tyler DJ, Triolo RJ. Optimization of selective stimulation parameters for multi-contact electrodes. Journal of Neuroengineering and Rehabilitation. 2013;**10**(1):25

[6] Schiefer MA, Freeberg M, Pinault G, Anderson J, Hoyen H, Tyler D, Triolo R. Selective activation of the human tibial and common peroneal nerves with a flat interface nerve electrode. Journal of Neural Engineering. 2013;**10**(5):056006

[7] Raspopovic S, Capogrosso M, Petrini FM, Bonizzato M, Rigosa J, Di Pino G, Carpaneto J, Controzzi M, Boretius T, Fernandez E, Granata G, Oddo CM, Citi L, Ciancio AL, Cipriani C, Carrozza MC, Jensen W, Guglielmelli E, Stieglitz T, Rossini PM, Micera S. Restoring natural sensory feedback in real-time bidirectional hand prostheses. Science Translational Medicine. 2014;**6**(222):222ra219

[8] Tan DW, Schiefer MA, Keith MW, Anderson JR, Tyler J, Tyler DJ. A neural interface provides long-term stable natural touch perception. Science Translational Medicine. 2014;**6**(257):257ra138

[9] Brenner E, Smeets JB. Size illusion influences how we lift but not how we grasp an object. Experimental Brain Research. 1996;**111**(3):473-476

[10] Flanagan JR, Bowman MC, Johansson RS. Control strategies in object manipulation tasks. Current opinion in Neurobiology. 2006;**16**(6):650-659

[11] Samuelsson K, Wressle E. User satisfaction with mobility assistive devices: An important element in the rehabilitation process. Disability and Rehabilitation. 2008;**30**(7):551-558

[12] Marasco PD, Kim K, Colgate JE, Peshkin MA, Kuiken TA. Robotic touch shifts perception of embodiment to a prosthesis in targeted reinnervation amputees. Brain. 2011;**134**(Pt 3):747-758

[13] Murray CD. Embodiment and prosthetics. Psychoprosthetics, Springer. 2008. pp. 119-129

[14] Caspar EA, Cleeremans A, Haggard P. The relationship between human agency and embodiment. Conscious and Cognition. 2015;**33**:226-236

[15] Robles-De-La-Torre G. The importance of the sense of touch in virtual and real environments. Ieee Multimedia. 2006;**13**(3):24-30

[16] Haggard P, Clark S, Kalogeras J. Voluntary action and conscious awareness. Nature Neuroscience. 2002;**5**(4):382-385

[17] Hogan N. The mechanics of multi-joint posture and movement control. Biological Cybernetics. 1985;**52**(5):315-331

[18] Kawato M. Internal models for motor control and trajectory planning. Current Opinion in Neurobiology. 1999;**9**(6):718-727

[19] Sigrist R, Rauter G, Riener R, Wolf P. Augmented visual, auditory, haptic, and multimodal feedback in motor learning: a review. Psychonomic Bulletin & Review. 2013;**20**(1):21-53

[20] Winters JM, Crago PE Biomechanics and Neural Control of Posture and Movement. Springer-Verlag New York Inc., New York, NY, USA; 2000

[21] Ajiboye AB, Weir RF. A heuristic fuzzy logic approach to EMG pattern recognition for multifunctional prosthesis control. IEEE Transactions on Neural Systems and Rehabilitation Engineering. 2005;**13**(3):280-291

[22] Hassan M, Kadone H, Suzuki K, Sankai Y. Wearable gait measurement system with an instrumented cane for exoskeleton control. Sensors. 2014;**14**(1):1705-1722

[23] Kiguchi K, Tanaka T, Fukuda T. Neuro-fuzzy control of a robotic exoskeleton with EMG signals. IEEE Transactions on Fuzzy Systems. 2004;**12**(4):481-490

[24] Veneman JF, Kruidhof R, Hekman EE, Ekkelenkamp R, Van Asseldonk EH, Van Der Kooij H. Design and evaluation of the LOPES exoskeleton robot for interactive gait rehabilitation. Ieee Transactions on Neural Systems Rehabilitation. 2007;**15**(3):379-386

[25] Roetenberg D, Baten CT, Veltink PH. Estimating body segment orientation by applying inertial and magnetic sensing near ferromagnetic materials. IEEE Transactions on Neural Systems Rehabilitation Engineering. 2007;**15**(3):469-471

[26] Mackey JR, Davis BL. Simultaneous shear and pressure sensor array for assessing pressure and shear at foot/ground interface. Journal of Biomechanics. 2006;**39**(15):2893-2897

[27] Kawasaki H, Komatsu T, Uchiyama K. Dexterous anthropomorphic robot hand with distributed tactile sensor: Gifu hand II. IEEE/ASME Transactions on Mechatronics. 2002;**7**(3):296-303

[28] Collins SH, Kuo AD. Recycling energy to restore impaired ankle function during human walking. PLoS One. 2010;**5**(2):e9307

[29] Nataraj R, Audu ML, Triolo RJ. Center of mass acceleration feedback control of functional neuromuscular stimulation for standing in presence of internal postural perturbations. Journal of Rehabilitattion Research and Development. 2012;**49**(6):889-911

[30] Farris RJ, Quintero HA, Murray SA, Ha KH, Hartigan C, Goldfarb M. A preliminary assessment of legged mobility provided by a lower limb exoskeleton for persons with paraplegia. IEEE Transactions on Neural Systems Rehabilitation Engineering. 2014;**22**(3):482-490

[31] Schofield JS, Evans KR, Carey JP, Hebert JS. Applications of sensory feedback in motorized upper extremity prosthesis: a review. Expert Review of Medical Devices. 2014;**11**(5):499-511

[32] Dhillon GS, Horch KW. Direct neural sensory feedback and control of a prosthetic arm. IEEE Transactions on Neural Systems Rehabilitation Engineering. 2005;**13**(4):468-472

[33] Marasco PD. Restoring Upper Limb Movement Sense to Amputees; A Move Towards Natural Contro. National Institutes of Health - NINDS; 2013

[34] Okamura AM, Dennerlein JT, Howe RD. Vibration feedback models for virtual environments. Robotics and Automation. 1998. Proceedings. 1998 IEEE International Conference on, IEEE; 1998, pp. 674-679

[35] Chiari L, Dozza M, Cappello A, Horak FB, Macellari V, Giansanti D. Audio-biofeedback for balance improvement: an accelerometry-based system. IEEE Transactions on Biomedical Engineering. 2005;**52**(12):2108-2111

[36] Nataraj R, Van den Bogert AJ. Simulation Analysis of Linear Quadratic Regulator Control of Gait. 40th Annual Meeting of the American Society of BiomechanicsRaleigh. USA: North Carolina; 2016

[37] Ackermann M, van den Bogert AJ. Optimality principles for model-based prediction of human gait. Journal of Biomechanics. 2010;**43**(6):1055-1060

[38] van den Bogert AJ. 2011, http://hmc.csuohio.edu/resources/musculoskeletal-modeling-and-simulation

[39] van den Bogert AJ, Blana D, Heinrich D. Implicit methods for efficient musculoskeletal simulation and optimal control. Procedia IUTAM. 2011;**2**(2011):297-316

[40] Varga A. On solving discrete-time periodic Riccati equations. Proc. of 16th IFAC World Congress, Prague, July 3-8; 2005

[41] Nataraj R, Audu ML, Kirsch RF, Triolo RJ. Comprehensive joint feedback control for standing by functional neuromuscular stimulation-a simulation study. IEEE Transactions on Neural Systems Rehabilitation Engineering. 2010;**18**(6):646-657

[42] Nataraj R, Audu ML, Triolo RJ. Comparing joint kinematics and center of mass acceleration as feedback for control of standing balance by functional neuromuscular stimulation. Journal of Neuroengineering Rehabilitation. 2012;**9**:25

[43] Hanspal R, Fisher K, Nieveen R. Prosthetic socket fit comfort score. Disability and rehabilitation. 2003;**25**(22):1278-1280

[44] Ehrsson HH, Rosén B, Stockselius A, Ragnö C, Köhler P, Lundborg G. Upper limb amputees can be induced to experience a rubber hand as their own. Brain. 2008;**131**(12):3443-3452

[45] Velliste M, Perel S, Spalding MC, Whitford AS, Schwartz AB. Cortical control of a prosthetic arm for self-feeding. Nature. 2008;**453**(7198):1098-1101

[46] van der Wel RP, Sebanz N, Knoblich G. The sense of agency during skill learning in individuals and dyads. Conscious Cognition. 2012;**21**(3):1267-1279

[47] Todorov E, Erez T, Tassa Y. MuJoCo: A physics engine for model-based control," Intelligent Robots and Systems (IROS). 2012 IEEE/RSJ International Conference on, IEEE; 2012. pp. 5026-5033

[48] Hebert JS, Olson JL, Morhart MJ, Dawson MR, Marasco PD, Kuiken TA, Chan KM. Novel targeted sensory reinnervation technique to restore functional hand sensation after transhumeral amputation. IEEE Transactions on Neural Systems Rehabilitation Engineering. 2014;**22**(4):765-773

[49] Li K, Nataraj R, Marquardt TL, Li ZM. Directional coordination of thumb and finger forces during precision pinch. PLoS One. 2013;**8**(11):e79400

[50] Nataraj R, Audu ML, Li ZM. Digit mechanics in relation to endpoint compliance during precision pinch. Journal of Biomechanics. 2015;**48**(4):672-680

[51] Moore JW, Obhi SS. Intentional binding and the sense of agency: A review. Conscious Cognition. 2012;**21**(1):546-561

[52] Hargrove L, Englehart K, Hudgins B. The effect of electrode displacements on pattern recognition based myoelectric control. Conference Proceedings IEEE Engineering Medicine and Biology Society. 2006;**1**:2203-2206

[53] Hargrove L, Losier Y, Lock B, Englehart K, Hudgins B. A real-time pattern recognition based myoelectric control usability study implemented in a virtual environment. Conf Proceedings IEEE Engineering Medicine and Biology Society. 2007;**2007**:4842-4845

[54] Earley EJ, Hargrove LJ, Kuiken TA. Dual window pattern recognition classifier for improved partial-hand prosthesis control. Frontiers Neuroscience. 2016;**10**:58

[55] Chan AD, Englehart KB. Continuous myoelectric control for powered prostheses using hidden Markov models. IEEE Transactions on Biomedical Engineering. 2005;**52**(1): 121-124

[56] Chan FH, Yang YS, Lam FK, Zhang YT, Parker PA. Fuzzy EMG classification for prosthesis control. IEEE Transactions Rehabilitattion Engineering. 2000;**8**(3):305-311

[57] Englehart K, Hudgins B, Parker PA, Stevenson M. Classification of the myoelectric signal using time-frequency based representations. Medical Engineering and Physics. 1999;**21**(6-7):431-438

[58] Hargrove LJ, Li G, Englehart KB, Hudgins BS. Principal components analysis preprocessing for improved classification accuracies in pattern-recognition-based myoelectric control. IEEE Transactions on Biomedical Engineering. 2009;**56**(5):1407-1414

[59] Simon AM, Hargrove LJ, Lock BA, Kuiken TA. Target achievement control 'test: Evaluating real-time myoelectric pattern-recognition control of multifunctional upper-limb prostheses. Journal of Rehabilitation and Research Development. 2011;**48**(6):619-627

[60] Kobrinski A, Bolkovitin S, Voskoboinikova L, Ioffe D, Polyan E, Popov B, Slavutski YL, Sysin AY, Yakobson YS. Problems of bioelectric control. Automatic and Remote Control, Proceedings of 1stIFAC International Congress. 1960;**2**:619

[61] Lee SW, Wilson KM, Lock BA, Kamper DG. Subject-specific myoelectric pattern classification of functional hand movements for stroke survivors. IEEE Transactions on Neural Systems and Rehabilitation Engineering. 2011;**19**(5):558-566

[62] McCloskey D. Differences between the senses of movement and position shown by the effects of loading and vibration of muscles in man. Brain Research. 1973;**61**:119-131

[63] Lock B, Englehart K, Hudgins B. Real-time myoelectric control in a virtual environment to relate usability vs. accuracy. Myoelectric Symposium. Proceedings of the MEC'05 conference, Fredericton, New Brunswick, Canada; 2005

[64] Crowninshield RD, Brand RA. A physiologically based criterion of muscle force prediction in locomotion. Journal of Biomechanics. 1981;**14**(11):793-801

[65] Van der Helm F, Veeger H. Quasi-static analysis of muscle forces in the shoulder mechanism during wheelchair propulsion. Journal of Biomechanics. 1996;**29**(1):39-52

[66] Boninger ML, Towers JD, Cooper RA, Dicianno BE, Munin, MC. Shoulder imaging abnormalities in individuals with paraplegia. Journal of Rehabilitation and Research Development. 2001;**38**(4):401-408

[67] van Drongelen S, van der Woude LH, Veeger HE. Load on the shoulder complex during wheelchair propulsion and weight relief lifting. Clinical Biomechanics (Bristol, Avon). 2011;**26**(5):452-457

[68] Veeger HE, Rozendaal LA, van der Helm FC. Load on the shoulder in low intensity wheelchair propulsion. Clinical Biomechanics (Bristol, Avon). 2002;**17**(3):211-218

[69] Nawoczenski DA, Riek LM, Greco L, Staiti K, Ludewig PM. Effect of shoulder pain on shoulder kinematics during weight-bearing tasks in persons with spinal cord injury. Archives of Physical Medicine Rehabilitation. 2012;**93**(8):1421-1430

[70] Finley MA, Rasch EK, Keyser RE, Rodgers MM. The biomechanics of wheelchair propulsion in individuals with and without upper-limb impairment. Journal of Rehabilitation and Research Development. 2004;**41**(3B):385-395

[71] Reinold MM, Escamilla RF, Wilk KE. Current concepts in the scientific and clinical rationale behind exercises for glenohumeral and scapulothoracic musculature. Journal of Orthopaedic and Sports Physical Therapy. 2009;**39**(2):105-117

[72] Kankipati P, Boninger ML, Gagnon D, Cooper RA, Koontz AM. Upper limb joint kinetics of three sitting pivot wheelchair transfer techniques in individuals with spinal cord injury. Journal of Spinal Cord Medicine. 2014;**38**(4):485-497

[73] McMonagle JS, Vinson E. MRI of the shoulder: Rotator cuff. Applied Radiology. 2012; **41**(4):8

[74] Fleckenstein JL, Canby RC, Parkey RW, Peshock RM. Acute effects of exercise on MR imaging of skeletal muscle in normal volunteers. AJR American Journal of Roentgenology. 1988;**151**(2):231-237

[75] Maas SA, Ellis BJ, Ateshian GA, Weiss JA. FEBio: finite elements for biomechanics. Journal of Biomechanical Engineering. 2012;**134**(1):011005

Chapter 5

Passive Biomimetic Prosthesis

Smita Nayak and Prasanna Lenka

Additional information is available at the end of the chapter

http://dx.doi.org/10.5772/intechopen.72123

Abstract

Advancement in prosthetic technology provides the prosthesis more natural function and cosmesis to the amputee limb. But in most of the cases due to some limitations of the patient, the most appropriate prosthetic solution is not possible. Custom made silicone prosthesis with some passive function plays the major role to rehabilitate those patients. The cosmetic prosthesis mimics the natural colour and texture of the normal body. Nowaday the cosmetic prosthesis becomes the foremost choice of the amputee individuals.

Keywords: prosthesis, cosmetic, passive, custom made, silicone

1. Introduction

Prosthesis (plural prostheses), the proper name for an artificial limb, derives from the Greek roots meaning "to replace an addition". The replacement of lost part was for the functional, cosmetic or protective reasons or some combinations of these. The earliest concept of cosmesis was started, and the first cosmetic wooden prosthesis of Hallux fitted in Egypt about 1000 BCE [1]. In the process of development and continuous research, the robotic prostheses are the new-generation prosthesis that mimics most natural movements and cosmesis. Advanced functional prostheses are with some disadvantages like high maintenance and cost and not reachable to every individual living in a rural area. The deformed shape of the stump in case of finger, partial hand and partial foot sometimes are not suitable for the fitment of functional prosthesis, but the patients are willing to fit the prosthesis to get cosmesis as well as some function. The passive prosthesis is the only option to fit any kind of irregular stump to make it cosmetic and provide some passive function. The individuals with passive prosthesis actively use their prostheses as frequently as do functional prostheses. The passive prostheses are used to stabilize objects, push against item and assist them in various ADL activities.

1.1. Cosmetic silicone prosthesis

In the development of material science, the silicone is found to be the most acceptable material to fabricate the cosmetic prosthesis. The silicone prosthesis can often restore a near-normal function in distal phalange amputations. The fabrication of the prosthesis required an artist hand with implementation of scientific knowledge [2]. The role of prosthesis is to replace the part by providing natural function and shape. The psychological aspect is more important to keep in mind during fabrication of any prosthesis. Aesthetic silicone prostheses have been shown to play a useful role in restoring normal appearance and assist in the rehabilitation of patients with amputations involving the upper limbs. Custom made finger prosthesis is aesthetically acceptable and comfortable for use in patients with amputated fingers, resulting in psychological improvement and well-being [3] (**Figures 1** and **4**).

The aesthetic hand prosthesis is previously used for cosmetic purposes, but the functional aspect has not been considered [4]. But role of this prosthesis is found in both cosmesis and passive functions (**Figure 2**). The usefulness of aesthetic prostheses is confirmed by the improvement of patients' living conditions and the continued wearing of these prostheses by the patients. Although a variety of materials have been used for aesthetic restoration, silicone is generally preferred because of its versatility, durability and compatibility with human tissue. Polymers of dimethyl siloxane (silicones) allow copying of the natural hand in every detail. Silicone prostheses are usually of high quality, match well with the patients remaining digits and thus are more aesthetically pleasing; long-term wearing of this prosthesis confirms as therapeutic tools. To achieve the normal appearance in the prosthesis, the cosmetic nail, the hair and flocking are used.

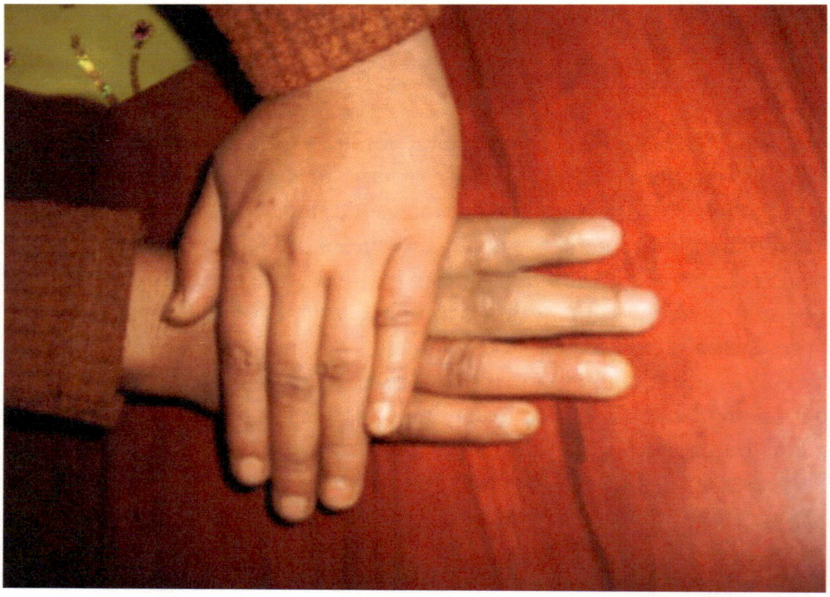

Figure 1. After fitment of silicone finger prosthesis in right side middle and index finger amputation.

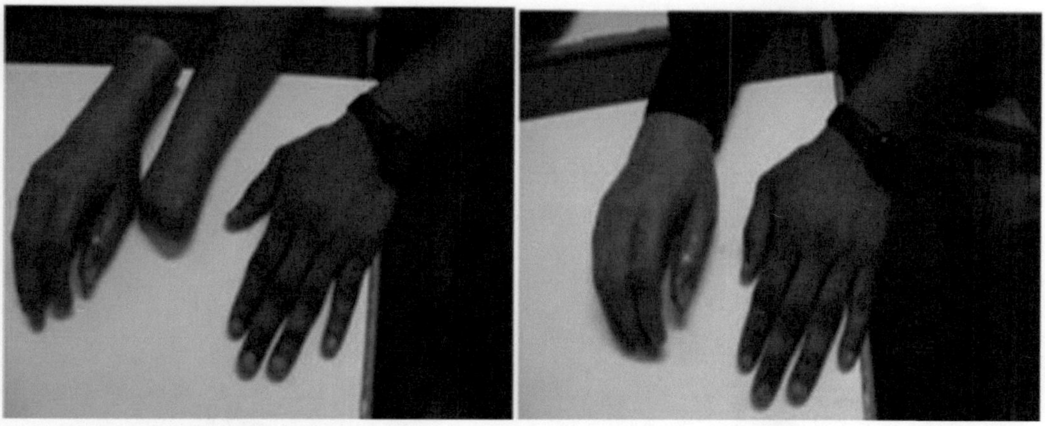

Figure 2. Partial silicone hand prosthesis.

Apart from the finger and hand, toe and partial foot silicone prosthesis is mostly accepted by the patients [5] (**Figures 3, 5**). The cosmetic prosthesis mimics the natural appearance without any complicated mechanism. These are easy to use with very low maintenance. So, nowadays, the use of cosmetic prosthesis is more than a complicated functional prosthesis.

1.2. Cosmetic silicone prosthesis with passive function (custom made)

The cosmetic prosthesis of the finger and hand is not usually having the movable digits. The prosthesis is only used for the cosmesis purpose and psychological benefit for the patients. Now these prostheses are having some passive movements of the fingers due to the involvement of some copper wires inside the prosthesis during the time of fabrication. This process is cost-effective and time-saving [6] (**Figure 6**).

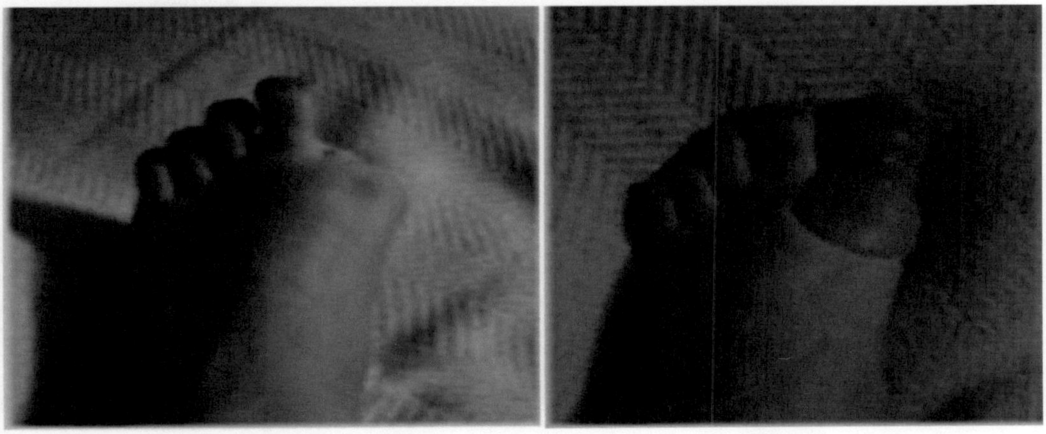

Figure 3. Before and after fitment of silicone toe.

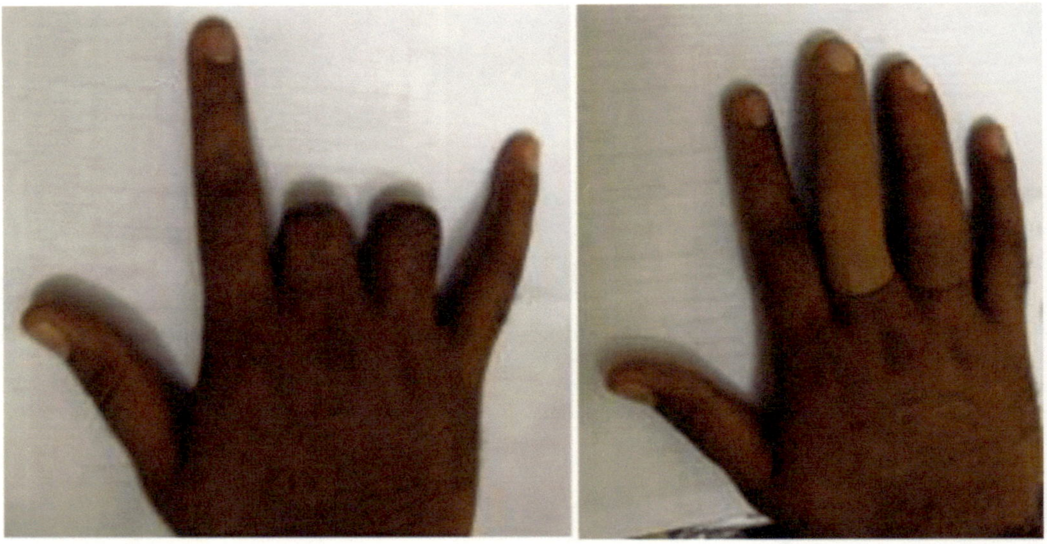

Figure 4. Fitting of silicone finger prosthesis in the middle and ring fingers.

1.3. Attachment and alignment of cosmetic prosthesis

The attachment of prosthesis depends upon the remaining part and contours of the stump. The finger amputation of PIP and DIP level, the suction suspension is the method of attachment, and amputation through metacarpophalangeal attachment is through strap suspension (**Figure 7**). The alignment is done before the fabrication during this process, and the overall length is checked with the opposite limb.

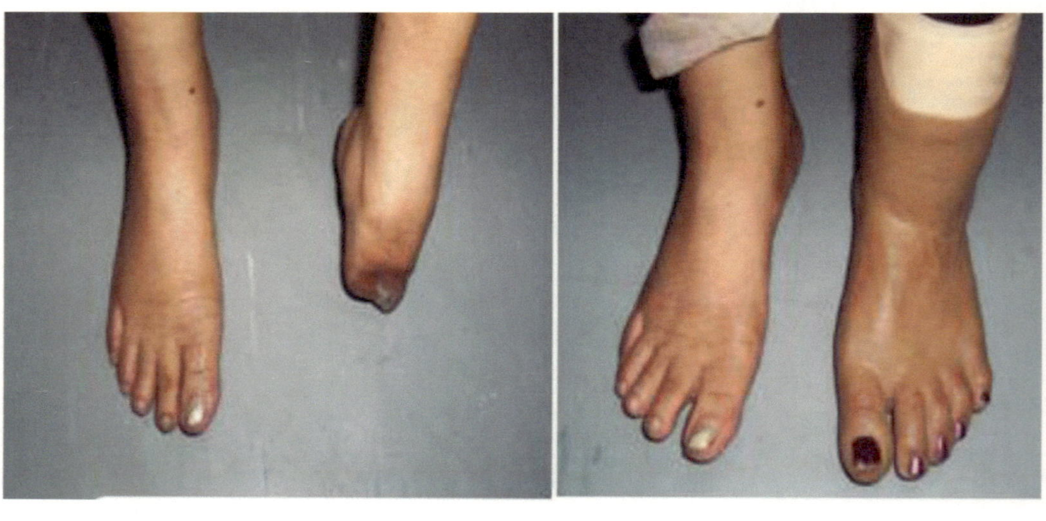

Figure 5. Silicone partial foot prosthesis.

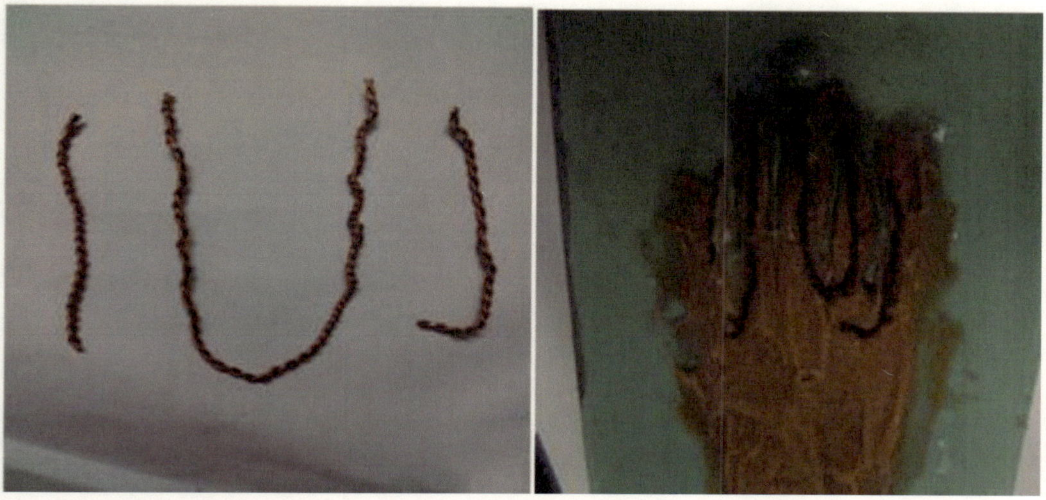

Figure 6. Copper wire and silicone die of partial hand with copper wire.

1.4. Maxillofacial silicone prosthesis

The face is considered as the most beautiful and expressive part of the body. The eye, ear, chick and nose, play different roles in various expressions (**Figures 8–11**). If the patient lost any part of the organ in the face, they felt as a half-dead person. These kinds of people were not having the courage to face the society. In the process of plastic surgery, they may modify the structure but

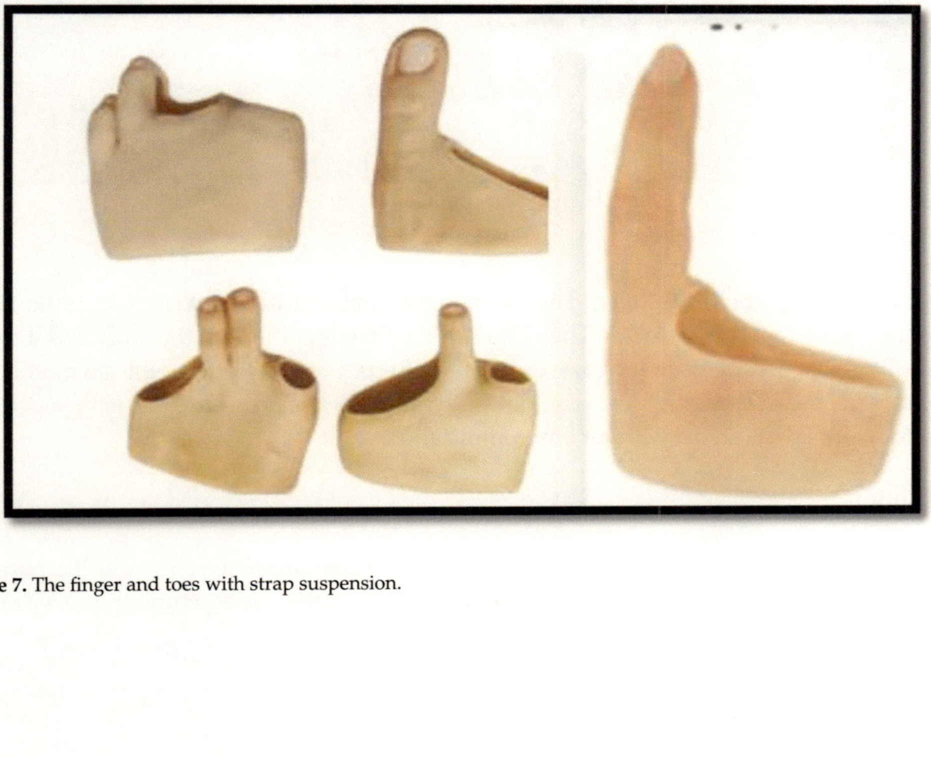

Figure 7. The finger and toes with strap suspension.

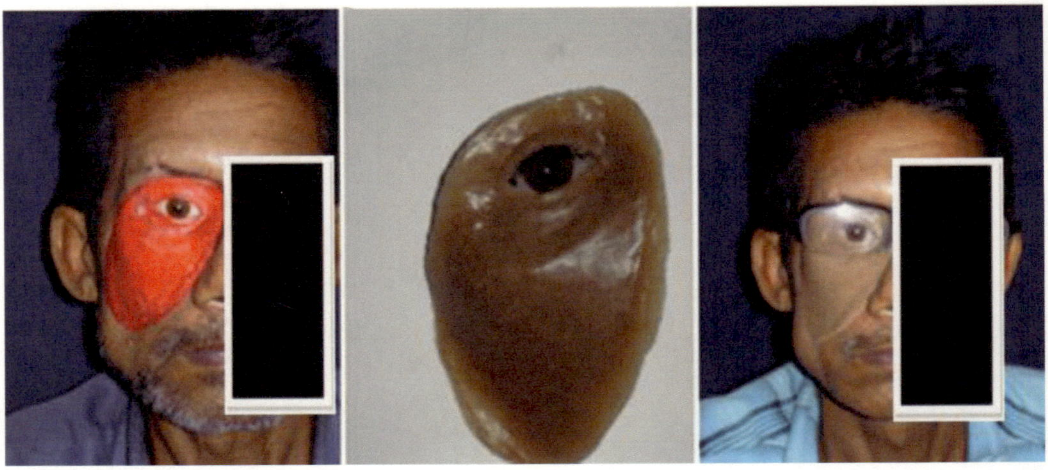

Figure 8. Patient with cosmetic eye.

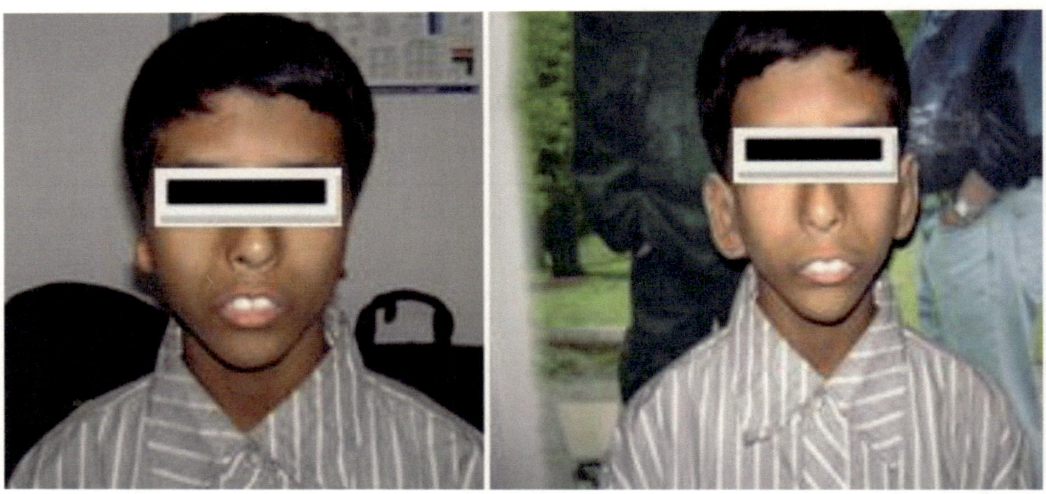

Figure 9. Before and after photo of cosmetic ear.

the scar and lost part not be fulfilled. The role of rehabilitation can be able to reduce the disfigurement by the cosmetic prosthesis. The silicone is very compatible with the body and it can be used as laryngeal prosthesis in the case of patient with tracheotomy during the chemotherapy. The number of breast cancer is increasing nowadays; the silicone breast is considered as the best option to maintain the shape without any complications.

1.5. Methods

The patients for passive prosthesis are selected according to the requirement of the patients, availability of the material and condition of the stump. The patients sometimes prefer cosmesis

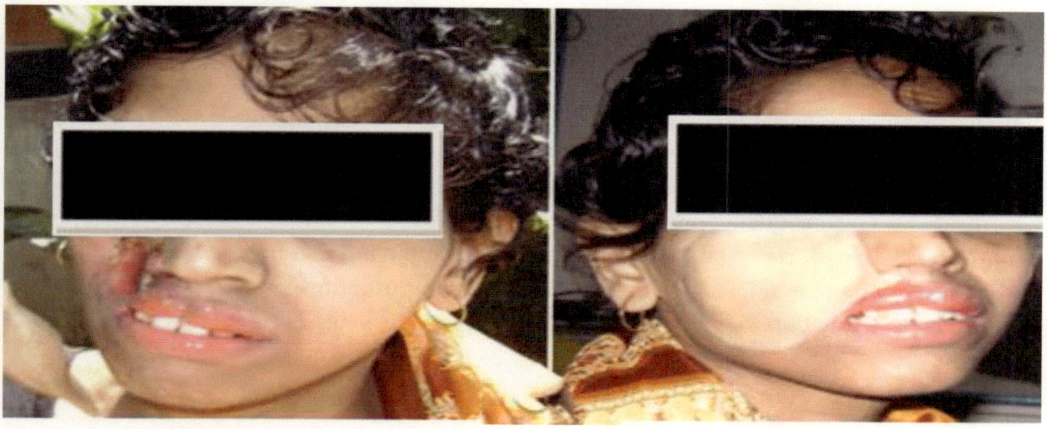

Figure 10. Before and after the use of silicone chick prosthesis.

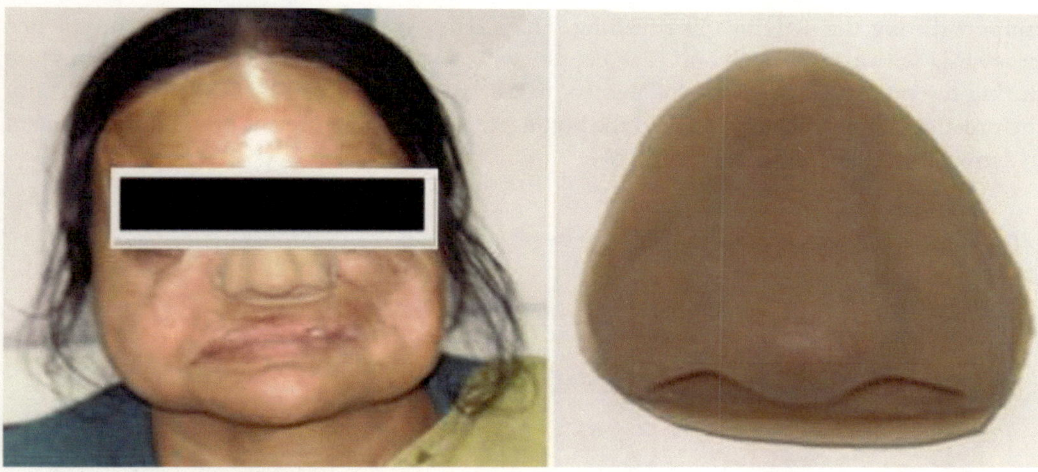

Figure 11. Patient with cosmetic silicone nose.

over functional prosthesis in that case the passive prosthesis plays the role of cosmetic prosthesis with the removal of additional component of functional prosthesis ultimately reduced the weight. The patients were assessed by the clinical team members and decided to fit the passive prosthesis, which was the suitable option for the patients. The criteria for selection of the patients were (1) uncosmetic appearance of the stump, (2) the functional prosthesis cannot be fitted, (3) the patient needs only cosmetic appearance, (4) unavailability of the functional components and (5) pain and insensitive stump. The selected patients were again assessed for the appropriate passive prosthesis such as finger prosthesis, partial hand prosthesis, toe prosthesis, below elbow prosthesis or maxillofacial prosthesis. The custom made passive prostheses were fabricated by using silicone material, which is the most preferable material nowadays having biocompatibility properties and most natural appearance [7–9]. The passive custom made prosthesis was fabricated in the silicone laboratory and fitted to the individual patients. The fabrication processes were the same for all the passive prostheses, but the casting and mould modification differ. The processes of fabrication were casting and measurement, mould modification, wax alignment, die preparation, colour matching, pouring, die compression and finishing of the prosthesis. The overview of the fabrication procedure of finger and hand is shown in **Figure 12**. The passive movements of the finger were achieved by the placement of copper wire inside the silicone during preparation of the die (**Figure 7**). Similar methods were followed in maxillofacial prosthesis. The functionality of finger and hand prosthesis were conducted by DASH questionnaire and Jebson hand function test and found this prosthesis plays the role in enhancement of function though treated as passive prosthesis [6].

1.6. Case studies

1.6.1. Case study 1

The male of 35 years old reported to the clinic having one middle finger amputation of the LT hand. He is professionally an engineer, and the chief complaint was that he had lost his support during the activities like feeding. The stump was good to fit the cosmetic silicone prosthesis with passive function. The copper wire was inserted inside the silicone mixture during the pouring process for the passive movement of the prosthesis. After the fitment of the prosthesis, he can be able to hold the bottle, got support to write and found improvement in activities of daily living (**Figure 13**).

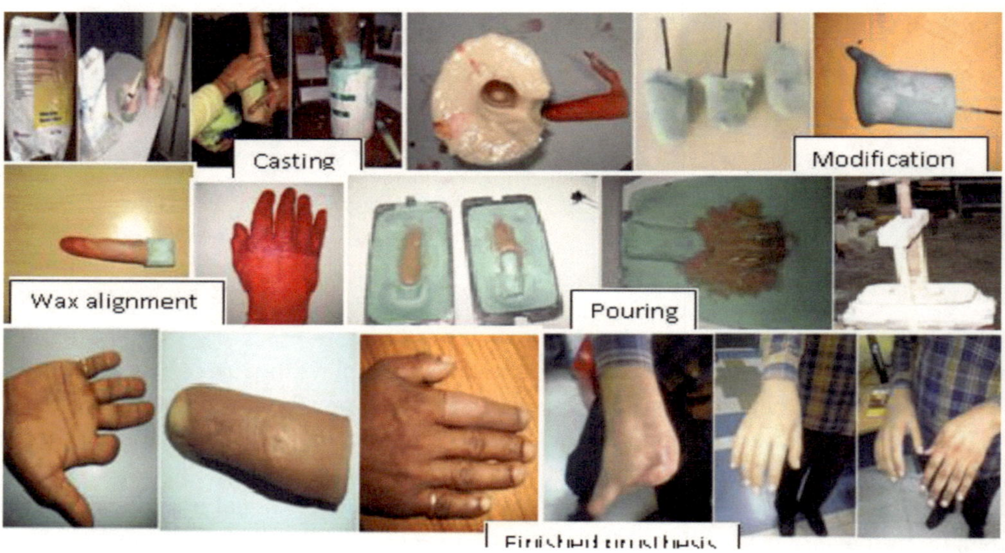

Figure 12. Procedure for fabrication of finger and hand prosthesis.

http://dx.doi.org/10.5772/intechopen.72123

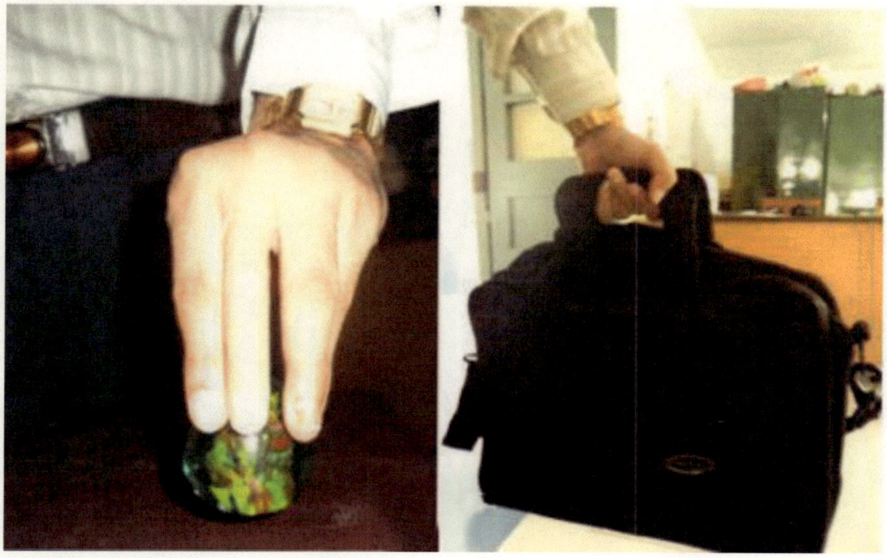

Figure 13. Patient with middle finger silicone prosthesis doing ADL.

1.6.2. Case study 2

A 12-year-old boy incurred with amputation of right first metacarpophalangeal joint (thumb) secondary to crush injury. He was a boy and a student, so the primary aim of fitting was supported during ADL's activities mainly during writing. The stump of the boy was not enough to fit the regular design with suction suspension. We decided to give him prosthesis with strap suspension. The boy was the donor for the opposite hand, and some changes in the alignment were required during the wax alignment process. For the strap the total cast of the hand was taken, and the area was decided in the hand where the maximum reduction can give to hold the finger in proper position. The prosthesis was fabricated and fitted successfully to get the support and cosmesis as required (**Figure 14**).

1.7. Components of passive cosmetic prosthesis

At present, cosmetic terminal devices consist largely of passive hands and various types of mechanical hands. It is therefore important to remember that cosmesis is attained only through the sacrifice of at least some degree of function. A woman may desire maximum cosmesis, preferring a soft, passive hand with a cosmetic glove, despite the fact that all functions are sacrificed. Different individuals may require different functions, but cosmesis is the priority for most of the users. The cosmetic silicone glove is an option, which can be used by all individuals with different occupations [10].

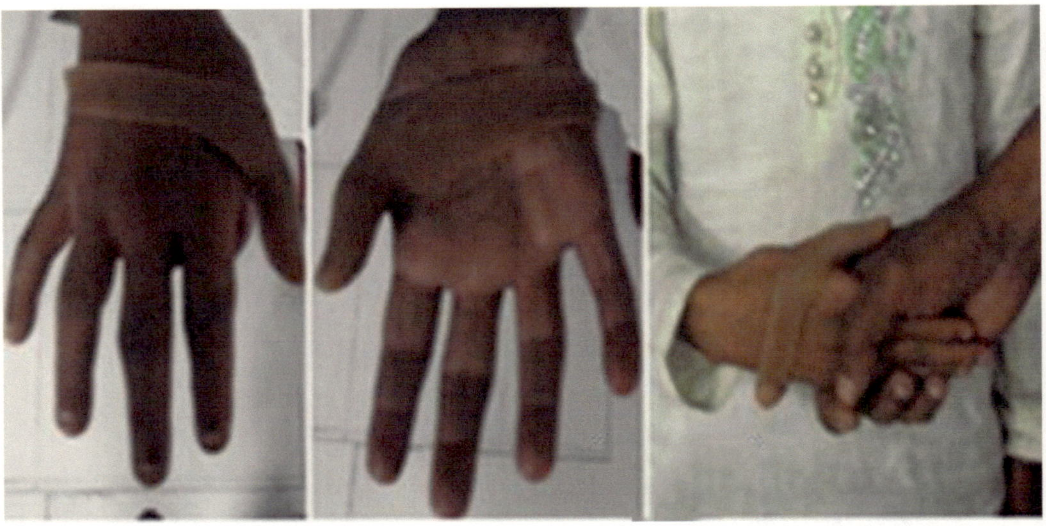

Figure 14. Cosmetic finger with strap suspension.

1.7.1. Cosmetic gloves

The gloves are made of polyvinyl chloride (PVC) plastic or advanced silicone and are made in moulds taken from human hands. As a result, all the structures like fingernails, lines, knuckles and vein prominences have a realistic appearance.

1.7.2. Glove colouring

The cosmetic glove can be of silicone and polyvinyl chloride varieties. The custom made varieties are matched maximum to the natural shape and colour of the body. The artificial hand colour remains unchanged in different climatic conditions. When choosing the colour, match the sound hand while it is hanging and in natural lighting. A colouring kit is used first to tone or "characterize" the colour of the glove. In the case of custom made silicone prosthesis during the time of colour matching, the required colour can be matched with the presence of the patient in midday time, and the colour of the nail can be fabricated by silicone or acrylic. The hair and vein can also be impregnated inside the mixed silicone to maintain the natural appearance of the prosthesis.

1.7.3. Passive hands

Passive hands are made of PVC foam over a flexible steel wire skeleton. The fingers and thumb can be positioned and repositioned by pressing the hand and bending the wire. Cosmetic gloves fit over the passive hands to give a natural looking. The passive hand is lighter than a mechanical hand and looks and feels more like human flesh. The passive hand, however, has no prehension. The passive hand is basically used for limited function in assisting the sound hand. Passive hands are also available without the wrist block; they can then be fitted directly to a socket, a forearm extension or a partial hand amputation.

1.7.4. Passive lower arms

Passive prosthesis for lower arm or below elbow amputation can be available in two forms, prefabricated or custom made. It is directly covered over the below elbow stump, and its colour is matched up to the expectation level of the amputee.

2. Discussion

The amputation of one or more fingers of the hand, as a consequence of trauma or congenital absence, carries a serious reduction in hand functional in addition to a psychological impairment [11, 12]. Jennifer Methot et al. (2010) studied grip strength in 50 subjects excluding ulnar two digits and found that the ulnar two digits also play a significant role in overall grip strength of the entire hand. Exclusion of the ulnar two digits resulted in a 34–67% decrease in grip strength, with a mean decrease of 55%. Exclusion of the little finger from a functional grip pattern decreased the overall grip strength by 33%. Exclusion of the ring finger from a functional grip pattern decreased the overall grip strength by 21%. It is clear that limitation of one or both of the ulnar digits adversely affects the strength of the hand. Custom made finger prosthesis is aesthetically acceptable and comfortable for use in patients with amputated fingers, resulting in psychological improvement and well-being [3, 4, 13]. The passive finger prosthesis in multiple finger amputation shows a significant improvement in cosmesis and function [12].

Analysis of function is an important component for hand rehabilitation. It assesses the initial limitations for the appropriate management. The hand function is the result of forceful flexion of all finger joints with the maximum voluntary force that the subject is able to exert under normal biokinetic condition. The synergistic action of flexor and extensor muscles and the interplay of muscle groups are important factors in the strength of hand [14].

The amputation of one finger of the hand, as a consequence of trauma, carries a serious reduction in hand function. The index finger plays a vital role in hand function strength [12]. According to Karle et al., the amputation of the proximal phalanx of the index finger found the loss of 17–35% in pinch strength [15]. Custom made finger prosthesis is aesthetically acceptable and comfortable for use in patients with amputated fingers, resulting in psychological improvement and well-being [3].

The functional improvement in partial hand shows after using the custom made silicone hand prosthesis in terms of ADL activities that the passive prosthesis is not only used for the cosmetic purpose but also helping in some passive functions. Pain was reduced markedly due to the compression provided by the prosthesis. Passive finger movement allowed the patient to ride his motorcycle and improved his ability to carry out different activities [6, 16]. The maxillofacial silicone prosthesis enhances the cosmesis and inner confidence of the patients to face the society [17].

Silicone elastomeric materials are more commonly used because they provide better stability and good marginal adaptation, which satisfies patient's cosmetic and aesthetic needs and possesses soft tissue-like consistency, and provide additional advantage when they are used to restore the defects in movable soft tissues [18].

Nowadays, the osseointegration of the ear with outer custom made silicone prosthesis was adopted by the patients though it integrated with the bone so the regular changing of prosthesis can be avoided [19–22]. Passive custom made tracheal prosthesis can be fabricated to provide the chemotherapy. The passive prosthesis plays the important role by providing comfort and cosmesis.

3. Conclusion

Cosmetic silicone prosthesis mimics the natural body structure, gives an aesthetic appearance and provides some passive functions. The cosmetic prosthesis are simple to use with low maintenance cost. The user can easily handle the prosthesis without any complication compared to high jagged prosthesis, but the maximum achievement of function is not possible by these prostheses. Apart from some limitations, these silicone cosmetic prostheses are found the most acceptable prosthesis nowadays and can able to retain the little smile on the face of the patients.

Acknowledgements

I would like to thank the National Institute for the Locomotor Disabilities in Kolkata, India, for providing the material and space for the study and the technical staff who actively participated in fabricating cosmetic prosthesis.

Author details

Smita Nayak[1]* and Prasanna Lenka[2]

*Address all correspondence to: smitank7@gmail.com

1 Pt. Deendayal Upadhaya National Institute for the Persons with Physically Disabilities (Divyangjan), New Delhi, India

2 National Institute for the Locomotor Disabilities (Divyangjan), Kolkata, India

References

[1] Wilson AB Jr. History of amputation surgery and prosthetics. In: Bowker JH, Michael JW editors. Atlas of Limb Prosthetics: Surgical, Prosthetic, and Rehabilitation Principles. 2nd ed. Rosemont, IL: American Academy of Orthopedic Surgeons; 2002. pp 3-15. (Originally published by Mosby- Year Book, 1992.)

[2] Taylor TD. Facial Prosthesis Fabrication Technical Aspects Clinical Maxillofacial Prosthesis. Chicago: Quintessence Publishing Company; 2000. p. 233

[3] Tripathi S, Singh DR, Chand P, Mishra N, Yadav LK, Singh V. A modified approach of impression technique for fabrication of finger prostheses. Prosthetics and Orthotics International. 2011;**36**(1):121-124

[4] Leow MEL, Pho RWH, Pereira BP. Esthetic prostheses in minor and major upper limb amputations. Hand Clinics. 2001;**17**:489-497

[5] Nayak S. Hand Book of Silicone Prosthesis. In: Editions Universitares Europeennes 2017. pp. 4-5 ISBN: 978-3-639-56067-1

[6] Nayak S, Lenka PK, Equebal A, Biswas A. Custom-made silicone hand prosthesis: A case study. Hand Surgery and Rehabilitation. 2016;**35**:299-303

[7] Pillet J. Esthetic hand prostheses. Journal of Hand Surgery. 1983;**8**:778-781

[8] Beasley RW. Hand and finger prostheses. Journal of Hand Surgery. 1987;**12A**:144-147

[9] Campbell GS, Gow D, Hooper G. Low cost cosmetic hand prostheses. Journal of Hand Surgery. 1992;**17B**:201-203

[10] Chapter 9 Cosmetic components. Retrieved Feb 20 2017. Available from www.cpousa.com/prosthetics/upper-extremity.

[11] Cervelli V, Bottini DJ, Arpino A, Grimaldi M, Rogliani M, Gentile P. Bone anchored implant in cosmetic finger reconstruction. Annales de chirurgie plastique esthetique. 2008;**53**:365-367

[12] Lifchez Scott D, Marchant-Hanson J, Matloub Hani S, Sanger James R, Dzwierzynski William W, Nguyen. Functional improvement with digital prosthesis use after multiple digit amputation. Journal of Hand Surgery. 2005 July;**30**(4):790-794

[13] Leow MEL, Ow RKK, Lee MH, Huak CY, Pho RWH. Assessment of color differences in silicone hand and digit prostheses: Perceptible and acceptable thresholds for fair and dark skin shades. Prosthetics and Orthotics International. 2006 April;**30**(1):5-16

[14] Radmin RG, Seoungyeon OH, Jenson TR, Webster JG. External finger forces in submaximal five finger static pinch Prehension. Ergonomics. 1992;**35**(3):275-288

[15] Karle B, Wittemann M, Germann G. Functional outcome and quality of life after ray amputation versus amputation through the proximal phalanx of the index finger. Handchir Mikrochir Plast chir. 2002;**34**(1):30-35

[16] Phillips SL, Harris MS, Latlief G. Experiences and outcomes with powered partial hand prostheses: A case series of subjects with multiple limb amputations. Journal of Prosthetics and Orthotics. 2012;**24**:93-97

[17] Gunay MEBGBKY. Facilitation of facial prosthesis placement with tattoo markers: A clinical report. Journal of Prosthetic Dentistry. 2007;**97**:265-260

[18] W. M. J. R. Aziza T. Analysis of the properties of silicone rubber maxillofacial prosthetic materials. Journal of Dentistry. 2003;**31**:67-74

[19] Wang S, Leng X, Zheng Y, Zhang D, Wu G. Prosthesis-guided implant restoration of an auricular defect using computed tomography and 3-dimensional photographic imaging

technologies: A clinical report. Journal of Prosthetic Dentistry. 2014; pii: S0022-3913(14)00425-9. DOI: 10.1016/j.prosdent.2014.08.014

[20] Hatamleh MM, Watson J. Construction of an implant-retained auricular prosthesis with the aid of contemporary digital technologies clinical report. Journal of Prosthodontics. 2013;**22**(2):132-136

[21] Kumar PS, Satheesh Kumar KS, Savadi RC. Bilateral implant-retained auricular prosthesis for a patient with congenitally missing ears A clinical report. Journal of Prosthodontics. 2012;**21**(2):322-327

[22] Todd JDA, Kubon M. An implant-retained auricular impression technique to minimize soft tissue. Journal of Prosthetic Dentistry. 2003;**89**:97-101

Chapter 6

Detection and Tracking of the Regions of Skin Using the Technique HS-ab

Diana Alejandra Contreras Alejo and
Francisco Javier Gallegos Funes

Additional information is available at the end of the chapter

http://dx.doi.org/10.5772/intechopen.70027

Abstract

The content of this work is the proposal of the technical HS-ab for the detection and tracking of the regions of skin in real time. First, two proposed techniques are analyzed for the modeling of skin color in images using a combination of color spaces HSV with YCbCr and HSV with CIELab. In the process of definition of the intervals of pixels is taken into account the following: non-skin color uses the components H and S, whereas skin color uses the components C_b, C_r, a, and b. The results showed that the HS-ab technique is better than the HS-CbCr technique because of the precision in detecting skin color according to the percentages C (34.8%) and CDR (67%). After, the morphologic operations are applied to debug the images of the previous segmentation and detect regions of skin using methods such as blob extraction and contour detection. Subsequently, the tracking of skin color consists of calculating the moments and positions of each frame to know the trajectory of the regions of skin. The purpose of the work is to design an easy-to-use computer vision system that will facilitate the early rehabilitation of patients before they are clinically ready to be fitted with a prosthesis.

Keywords: color spaces, skin, technique, threshold, tracking

1. Introduction

The loss of a limb in the body, such as the arms, hands, and legs, is one of the most devastating events that can happen to a person. Subsequently, people require an amputation in order to reduce disability, eliminate useless limbs, and save lives. The treatment of the amputated person includes not only the surgery but also the restoration of the function and the setting of an artificial limb. The treatment must be considered as a continuous dynamic process that begins at the time of injury and continues until the patient has reached the maximum usefulness of his or her prosthesis and is able to perform the essential activities of daily living. The treatment can be divided into two stages: the preprosthetic stage and the prosthetic stage. The first is rehabilitation before using the prosthesis, and the second is rehabilitation with the use of the prosthesis. Both stages consist of physical exercises to recover muscular strength, functional independence of the limb, and mobility of the residual limb [1]. This rehabilitation treatment can be carried out with computer vision systems such as the recognition system. This project develops a skin color recognition system so that it can be used with the residual limb and prosthesis of skin-colored frame.

Today, the use of recognition systems is increasingly being recognized as a useful tool for the study, assessment, and rehabilitation of functional abilities. This system can offer innovative and exciting ways to rehabilitate, making the treatment more enjoyable and, therefore, increasing the motivation of patients [2]. In addition, these systems have the ability to create a dynamic stimulus environment by offering an active behavioral response in patients, which cannot be obtained from traditional therapies. The applications of the system are focused on cognitive processes, including attention processes, spatial abilities, and memory. Application scenarios of recognition systems are designed to teach the basic activities of daily living such as common object recognition, meal preparation, shopping, etc. These applications allow patients a feedback when they continue their exercise program in the home environment [3].

The mission of this work is the development of applications focused on the rehabilitation of patients with amputation. On this occasion, this document presents the beginning of the project that is to detect and track the areas of skin on human body; therefore, the concepts that form a recognition system are described below.

The recognition systems such as face detection, corporal detection, hands tracking, etc. use skin detection as the main step in the system. Color image segmentation is useful in many applications, mainly for the purpose of human-computer interaction (HCI) [4, 5]. The segmentation results allow to identify regions of interest in different scenes, which is very beneficial to provide the needed information. In this work, the skin color is our interest, for which the segmentation of images in skin color is realized. The information in the color of the skin is useful for the detection, the location, and the tracking of the parts of human body; it also allows a fast processing and provides robustness in the application. A system of detection and tracking of skin consists of the following points: to choose an appropriate color space for the detection of skin color, to extract skin color by a modeling, to debug the regions of the skin and to follow the movements of the parts of skin in people. Each point in the system is described below.

1.1. Color spaces

The aim of skin color pixel classification is to determine if a color pixel is a skin color or non-skin color. A useful classification should differentiate all skin types as yellowish, brownish, whitish, etc., and stock different lighting conditions. The color classification is employed using only pixels chrominance because it is expected that skin segmentation may be more robust to lighting variations when the pixels luminance are discarded [6]. This section describes four color spaces, which are commonly used in the skin segmentation:

- RGB: colors are specified in three primary colors red (R), green (G), and blue (B). The advantage is simplicity but it does not separate luminance and chrominance.
- HSV: also known as HSB (B-brightness) color space. The component hue (H) defines the dominant color (such as red, green, purple, and yellow) of an area and varies from 0 to 1, saturation (S) measures the colorfulness of an area in proportion to its brightness, and value (V) is related to the color luminance, that is to say, it corresponds to brightness and it varies from 0 to 1. The HSV color space is computed using Eqs. (1)–(3). This model gives poor result where the brightness is very low. This color space discards luminance from chrominance. Other similar color spaces are hue, saturation, and intensity (HIS) and hue, saturation, and lightness (HSL).

$$H = \arccos \frac{½[(R-G)+(R-B)]}{\sqrt{[(R-G)^2+(R-B)(G-B)]}} \tag{1}$$

$$S = 1 - 3\frac{\min(R,G,B)}{R+G+B} \tag{2}$$

$$V = \frac{1}{3}(R+G+B) \tag{3}$$

- YCbCr: the Y, C_b, and C_r components refer to luminance, chromatic blue, and chromatic red. It provides an excellent color space for luminance and chrominance separability [7]. The simplicity of the transform in Eqs. (4)–(6) and the explicit separation of luminance and chrominance components make this color space very attractive for skin color detection [8].

$$Y = 0.299R - 0.587G - 0.114B \tag{4}$$

$$C_b = R - Y \tag{5}$$

$$C_r = B - Y \tag{6}$$

- CIELab: it is a perceptually uniform color space that was proposed by G. Wyszecki and standardized by Commission Internationale de L'Eclairage (CIE). It separates a luminance variable L from two perceptually uniform chromaticity variables a and b.

In this document, the color spaces HSV, YCrCb, and CIELab are used for the segmentation of color of the skin, due to the fact that they separate the components of chrominance and luminance achieving a characterization of the different colors of skin [9].

1.2. Skin color segmentation

A part of the process of the segmentation is to construct a rule of decision that discriminates or differs between the pixels of an image that corresponds to skin color and non-skin color. This can be done by nonparametric modeling as Bayesian classifier and histogram-based thresholding or parametric modeling such as Gaussian model (GM) and Gaussian mixture model (GMM) [10]. This paper describes the technique of histogram threshold to image segmentation of skin color. This technique is fast and practical in training, also it is theoretically independent from the shape of skin distribution [11].

The method uses an approximate estimate to define the interval of a color space corresponding to skin color that appears in the training images, see Eq. (7).

$$P_{skin}(c) = \frac{skin[c]}{Norm} \tag{7}$$

where $skin[c]$ is the interval value of the histogram, c corresponds to the color vector, and $Norm$ is the normalization coefficient (sum of all histogram values) or the maximum value of the intervals.

Some research [12, 13] used the histogram-based thresholding for segmenting the skin pixels that is projected to work with lighting conditions, skin tones, and for real-time skin segmentation in video.

1.3. Morphological operations

The detection system involves subsequently the usage of morphological operations to refine the skin regions extracted from the segmentation. Morphological operations are a set of simple local filters, which can be combined to obtain more complex results.

Two morphological operations are used in this chapter. Erosion, which reduces the image regions that represent non-skin color pixels, and dilation that expands the regions of the image of skin color pixels, which were lost due to the aggressive erosion applied in the previous step. Applying the erosion and later the dilation will allow to smooth contour of the objects, eliminate small protuberances, and disappear fine structures.

1.4. Detection and tracking

This section describes the procedures to detect and track the regions of skin in image sequences.

The first, the detection of constant changes can be carried out with the blob method, which it is possible to know the positions of the moving object [14]. A blob is a group of connected pixels in an image that shares some common property, for example, color. There are two commonly used manners for getting the blobs: one is to choose an appropriate threshold to differentiate moving objects (skin regions) from the background and another way is to conduct motion analysis [14]. The latter can be set using open source computer vision (OpenCV), which is a library of programming functions mainly aimed at real-time computer vision. There are three blob detection libraries for OpenCV, cvBlobsLib, cvBlob, and Bloblib that allow to know the positions of the blobs. On the other hand, in the first, the blobs are extracted (areas of skin color) with the threshold and create a binary mask of skin color pixels. Subsequently, the method of contour detection is applied, it has as main target to locate all the pixels that belong to the contour of the body parts, in other words, this stage detects all the contours that

http://dx.doi.org/10.5772/intechopen.70027

surround the blob that matches with the skin regions. This last form is used in this project, which is the extraction of blob with contour detection.

One way of tracking the skin regions is through the moments. In pattern recognition, the moments have been extensively used as global features of images [15]. The moments can calculate the position of the center of the skin regions. This calculation consists to calculate first-order spatial moments around x-axis and y-axis and the Zero-order central moments of the binary image. Zero-order central moments of the binary image are equal to the white area of the image in pixels. In the same way, this calculation can be carried out with OpenCV.

The chapter is organized as follows. Section 2 presents the procedure to calculate the threshold, the implementation of the techniques, the evaluation of the results of the techniques, and the description of the system algorithm. Section 3 demonstrates experimental results. Section 4 discusses the conclusions.

2. Methodology

The methodology is based on the following parts: determination of thresholds in the color spaces for skin color or non-skin color pixels, development of algorithm for the techniques HS-CbCr and HS-ab, evaluation of the results of the techniques and algorithm for detection, and tracking of skin color.

2.1. Segmentation of skin color and non-skin color with histogram based thresholding

Two techniques are proposed: HS-CbCr, which is the combination of the components H and S of HSV color space with the components C_b and C_r of the YCbCr color space, and the technique HS-ab (it is the same form of the technique previously) is the combination of H and S but with the components a and b of CIELab color space. To determine the intervals of pixels non-skin color uses the components H and S, in the case of the intervals of pixels skin color uses the components C_b, C_r, a, and b.

This section uses 80 images, of which 40 are images of different types of color skin obtained from the database of SFA [16] and the other 40 images are different colors to skin color.

In the case of non-skin images, each image is transformed to the HSV color space by obtaining the histograms of the channels H and S. In each, histogram is estimated the minimum and maximum values for the components H and S. After having all the values of the 40 images and using the Eq. (7), the minimum and maximum average values of H are obtained, as well as the minimum and maximum average values of S. In this way, the thresholds for non-skin color pixels are obtained.

In the case of skin images, it is the same procedure described as above except that it transforms into the YCbCr color space to obtain the thresholds of the components C_b and C_r and also converts into CIELab color space to get as result the thresholds of the components a and b. Then, to get the skin color pixel thresholds for each color space, only the chrominance components are considered.

For example, a non-skin image is transformed into the HSV color space, and the histograms of H and S are obtained. The histogram of H has a minimum value ($a1$) and a maximum value ($b1$), when applying to the 40 images, the minimum values are $a1 + a2 + \cdots + a40$, and the maximum values are $b1 + b2 + \cdots + b40$. Then, in Eq. (7), *skin*[*a*] is the average of all the minimum values of a while *Norm* is the maximum value of all values of *a*, this process is also performed for the maximum values so that the minimum and maximum thresholds of H are $P_{\text{skin}}(a) < H < P_{\text{skin}}(b)$. In this way, the intervals of the components of interest are obtained.

The procedure described in this section is shown in **Figure 1**. The algorithm for obtaining skin and non-skin color thresholds is implemented in MATLAB (R2016a—The MathWorks).

The aim of this section is to obtain the ranges of the components H and S for the non-skin color and the components C_b, C_r, a, and b for the skin color.

2.2. Techniques HS-CbCr and HS-ab

This chapter presents two techniques of color models for the detection of skin color. Both techniques use the HSV color space that provides additional information of hue and chrominance of an image with the aim to improve discrimination between skin pixels and non-skin pixels [17]. In this work, the HSV color space is used to detect non-skin color pixels later, in one technique, the YCbCr color space is applied to detect the skin color pixels and in the other, the CIELab color space.

After knowing the thresholds of each color space, an algorithm is developed to detect the skin regions of an image. The algorithm is implemented in Microsoft Visual Studio (C++ 2015). This algorithm consists of loading the image, transforming the image into HSV color space, and identifying non-skin pixels with the thresholds H and S to discard them. Later with the same image, it turns in the YCbCr color space to select the skin color pixels with the thresholds C_b and C_r and displays the image with only skin regions, which is known as the technique HS-CbCr. On the other hand, the technique HS-ab has the same procedure described earlier with

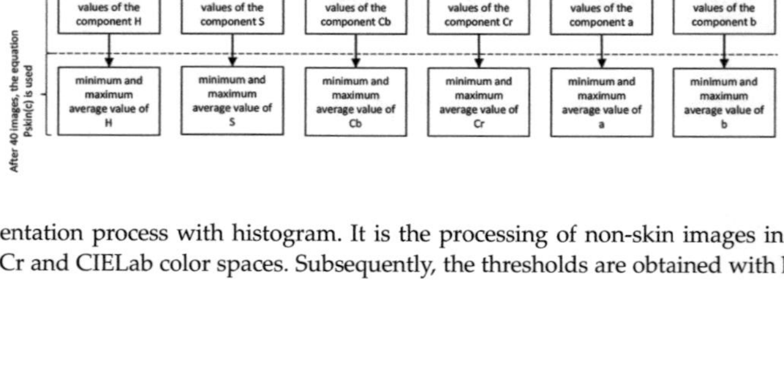

Figure 1. Color segmentation process with histogram. It is the processing of non-skin images in HSV color space and images of skin in YCbCr and CIELab color spaces. Subsequently, the thresholds are obtained with histograms.

http://dx.doi.org/10.5772/intechopen.70027

the exception that uses the thresholds *a* and *b* to identify skin color pixels. The algorithm of the techniques HS-CbCr and HS-ab is presented below:

```
Function SKIN SEGMENTATION
for all images C in the training set T do
Convert to HSV color space
for each pixel c in an image C do
if Pskin-min(h1)<H<Pskin-min(h2) and Pskin-min(s1)<S<Pskin-min(s2) then
Classify the pixel as non-skin color
Remove pixels in the image C
Convert to YCbCr/CIELab color space
for each pixel c in an image C do
if Pskin-min(Cb1/a1)<Cb/a<Pskin-min(Cb2/a2) and Pskin-min(Cr1/b1)
<Cr/b<Pskin-min(Cr2/b2) then
Classify the pixel as skin color
Remove pixels in the image C
Display image C
else
Re-read image C
end if
end for
else
Re-read image C
end if
end for
end for
end Function
```

The purpose of the development of the algorithm is to separate the skin areas against the non-skin areas in two different ways.

2.3. Performance evaluation

The metric parameters are used to evaluate the results of the skin detection algorithms. In this work, three different metrics, C, CDR, and FDR, are implemented as in [18, 19]. The C is the percentage of the correct skin detection to determine which chrominance component is the best skin detection result. The C is given in Eq. (8). The correct detection rate (CDR) is the percentage of the pixels that are classified correctly by the algorithm as skin pixels. It is expressed in Eq. (9). The false detection rate (FDR) is the percentage of the pixels that are classified wrongly by the algorithm as non-skin pixels. This corresponds to Eq. (10).

$$C = \frac{P}{T} \times 100\% \tag{8}$$

$$\mathrm{CDR} = \frac{P_c}{T_s} \times 100\% \tag{9}$$

$$\mathrm{FDR} = \frac{P_w}{T_{ns}} \times 100\% \tag{10}$$

Where P is the number of pixels correctly classified as skin color of some color space, T is the total number of pixels in the database (training images), P_c represents the total number of pixels that are classified correctly as skin pixels by the proposed algorithm, T_s is the total number of skin pixels that are classified by the ground truth, P_w represents the total number of pixels that are classified wrongly as non-skin pixels by the algorithm, and T_{ns} is the total number of non-skin pixels that are classified by the ground truth. The term ground truth refers to the information of the actual data, for example, the skin images are truthful information of skin color.

The performance metrics are essential concepts in verifying a specified algorithm as skin detection algorithm.

2.4. Skin color tracking

According to the results of the previous section, the best technique is selected to subsequently implement in the next algorithm. The algorithm is implemented in Microsoft Visual Studio (C++ 2015) with the library OpenCV 3.1.0, and this software is tested under Windows 10 of a computer ASUS Notebook UX32A. The algorithm is observed in **Figure 2** and consists of the following steps:

- **Step 1.** This step is to capture the image and to read temporary images via webcam, and then each frame image is converted into HSV to eliminate pixels of non-skin color and is subsequently transformed into the CIELab color space to represent only skin color pixels. Then, the values of the thresholds H, S, a, and b are implemented. Last, the image becomes binary, with 1 representing skin color pixels and 0 representing non-skin color pixels.
- **Step 2.** The application of morphological operations clarifies the regions of skin extracted from the segmentation of the previous step. This project applies erosion and dilation with the aim of separating areas of skin that are connected to areas not to skin that managed to survive the segmentation.
- **Step 3.** The methods of blob and contour detection are applied in this step. A blob is a connected part of a binary image that also refers to the regions in an image that are either brighter or darker than the surrounding. All blobs are classified in the image by their area,

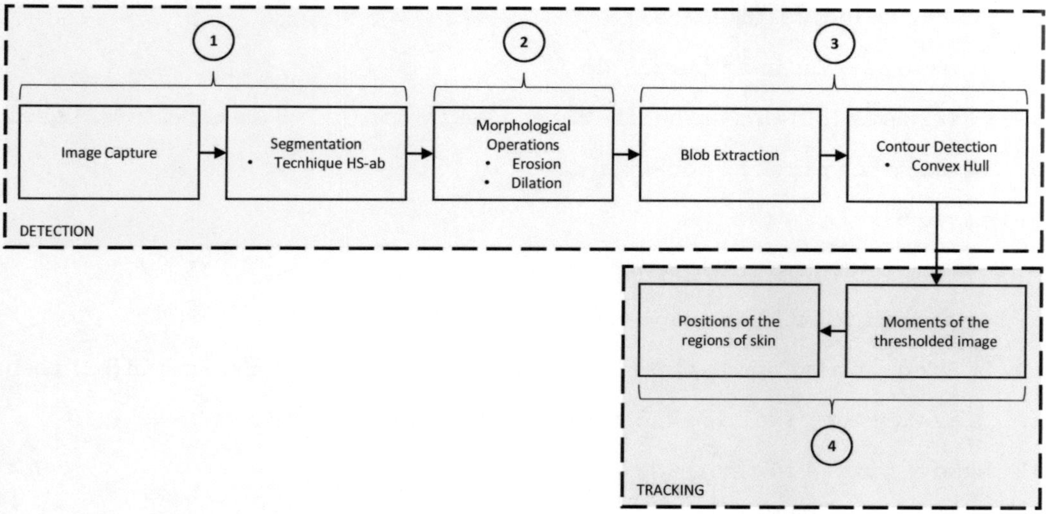

Figure 2. Detection and tracking of the areas of skin color in the image is a diagram on the development of the project.

then the background of the image is extracted, and the blob of the body parts is obtained. Later, contour detection is used to store all the contours of skin regions. There are two most common types, the convex hull and the ellipse. The first is elemental to get the convexities of the body gestures, while the second is defined as the rectangle that includes all the points that compose the contour of skin regions that is making a gesture. On this occasion, it works with the convex hull, although it is still not getting the information of the position of the parts of the body by each frame. However, it is a work for future to implement it at the stage of tracking. The library cvBlobsLib is used to generate and filter any blobs, and the function cvFindContours() that gives the contour (outline) and blob area. The library and the function are from OpenCV.

- **Step 4.** The tracking consists of knowing the positions of the user gesture to display on a computer screen the path of the movements. The problem is that the position of user can vary depending on the gesture performed. To carry out this, one must know the positions of the user's movement. The information of the positions is obtained through the moments. Moments moments() is an OpenCV function that calculates all of the spatial moments and returns a moments object with the results, and this function is applied. Then, it is considered that if the white area of the binary image is less than or equal to 10,000 pixels, then there are not skin zones in the image because the skin zones are expected to have an area more than 10,000 pixels. Therefore, the positions of the x-axis and y-axis in the image are calculated considering the moment between the areas. In this way, the path can be displayed in the continuous images.

The algorithm of the previously described steps is shown below:

```
1: Function SKIN SEGMENTATION
2: for all images C in the training set T do
```

```
Convert to HSV color space
for each pixel c in an image C do
if Pskin-min(h1)<H<Pskin-min(h2) and Pskin-min(s1)<S<Pskin-min(s2) then
Classify the pixel as non-skin color
Remove pixels in the image C
Convert to CIELab color space
for each pixel c in an image C do
if Pskin-min(a1)<a<Pskin-min(a2) and Pskin-min(b1)<b<Pskin-min(b2) then
Classify the pixel as skin color
Remove pixels in the image C
Display image C
Apply morphological operations
Extract blobs
Detect contour
Calculate the moments
if area<=10000 then
Calculate the positions
if (X1 >= 0 && Y1 >=0 && X2 >= 0 && Y2 >= 0) then
Draw the line
else
X1=X2, Y1=Y2
else
Re-calculate area
else
Re-read image C
end if
end for
else
Re-read image C
```

http://dx.doi.org/10.5772/intechopen.70027

```
end if
end for
Show continuous images
end for
end Function
```

This section explains the process of design of a computer vision system for tracking of the skin areas of the user. Then, the following section presents the results.

3. Results

In this section, representative results are provided for a proposed algorithm for the detection and tracking of human skin parts. As mentioned, each non-skin image is converted into HSV color space, and each skin image is transformed into YCbCr and CIELab color spaces to obtain the maximum and minimum thresholds of the channels H, S, C_b, C_r, a, and b. An example is presented in **Figure 3**, where a skin image is converted into the YCbCr color space and its respective histograms of each channel are presented where its thresholds are $C_b = [110, 114]$ and $C_r = [138, 142]$, which shows that this case is the procedure of an alone image. In this work, 40 skin images and 40 non-skin images are used to obtain the thresholds based on Eq. (7). For non-skin images, the thresholds of H and S are $0 < H < 0.2$ and $0.15 < S < 0.9$, whereas for skin images in YCbCr color space, the thresholds are $88 < C_b < 130$ and $127 < C_r < 175$, and in CIELab color space, the thresholds are $142 < a < 225$ and $115 < b < 177$. These thresholds are implemented in the algorithm of the techniques HS-CbCr and HS-ab.

For the application of the techniques HS-CbCr and HS-ab, 20 real images are used where people are able to detect areas of skin. After implementing the thresholds of H, S, C_b, C_r, a, and b in the algorithm of the techniques HS-CbCr and HS-ab, **Figure 4** presents the results of some real images with the application of both techniques.

According to the experimental results, it is possible to say that the images with the technique HS-ab present better results in comparison with the technique HS-CbCr, since the latter technique shows false alarms in the confusion of the colors of sand, wood, hair brown, among others such as skin, i.e., the colors similar to the skin are detected as skin. To obtain quantitative results, both techniques are evaluated with the metric parameters.

For the use of mathematical equations, Eqs. (8)–(10) are considered as follows:

P is the total number of skin color pixels of the skin images after transforming to the YCbCr and CIELab color space taking into account only the chrominance components, T is the total number of skin color pixels of the 40 skin images after converting to YCbCr and CIELab color space. In this case, two percentages of C are obtained, one for the YCbCr color space and another for CIELab. On the other hand, two percentages of CDR are obtained, one for the

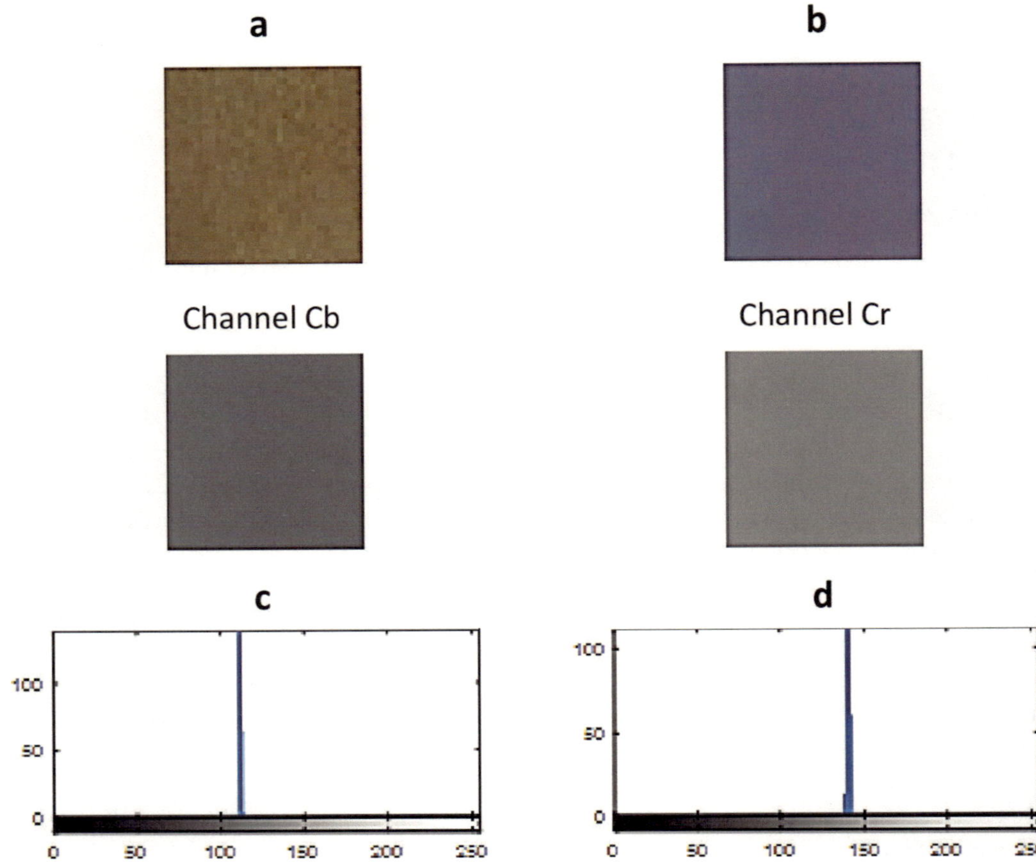

Figure 3. A skin image in YCbCr color space with histograms. In this, a is a skin image, b is the conversion to the YCbCr color space, c is the histogram of the channel C_b, and d is the histogram of the channel C_r. In both histograms, the minimum and maximum thresholds of the components C_b and C_r are obtained.

technique HS-CbCr and another for the technique HS-ab, where P_c is the total number of skin color pixels found in the algorithm HS-CbCr and HS-ab on real images, and T_s is the total number of skin color pixels of the 40 skin images. FDR has the same percentage for both techniques, since in detecting, the non-skin color pixels use the same procedure, then P_w is the total number of non-skin color pixels detected in the algorithms HS-CbCr and HS-ab on real images, and T_{ns} is the total number of non-skin color pixels of the 40 non-skin images. These percentages appear in **Table 1**.

As seen in **Table 1**, the percentage of C in CIELab is higher with respect to YCbCr, which means that the CIELab color space is more accurate in detecting skin color with chrominance components; therefore, the best results are expected if the CIELab color space is used. The proposed algorithm of the technique HS-ab presents a high detection of pixels correctly classified as skin color compared to the technique HS-YCbCr, due to the fact that HS-ab has a greater CDR than HS-CbCr besides not presenting false alarms. This is the reason for the choice of technique HS-ab for the application in the next algorithm.

Figure 4. Images with application of the proposed techniques. From left to right, in the first column the real images are shown, in the second column are the images with the application of the technique HS-CbCr, and in the latter the images are processed with the technique HS-ab.

Parameters	YCbCr color space	CIELab color space	Technique HS-CbCr	Technique HS-ab
C (%)	33.51	34.80	–	–
CDR (%)	–	–	64.88	67
FDR (%)	–	–	48.87	48.87

Table 1. Percentages of the three metric parameters.

The results of algorithm for detecting and tracking skin color are presented in **Figure 5**. The operation of the algorithm is to capture the continuous images through the camera of the PC, showing on the screen the images of blob and the path in red color of the regions of greater area of skin in motion.

This system has been tested with different skin types for which a satisfactory response is obtained.

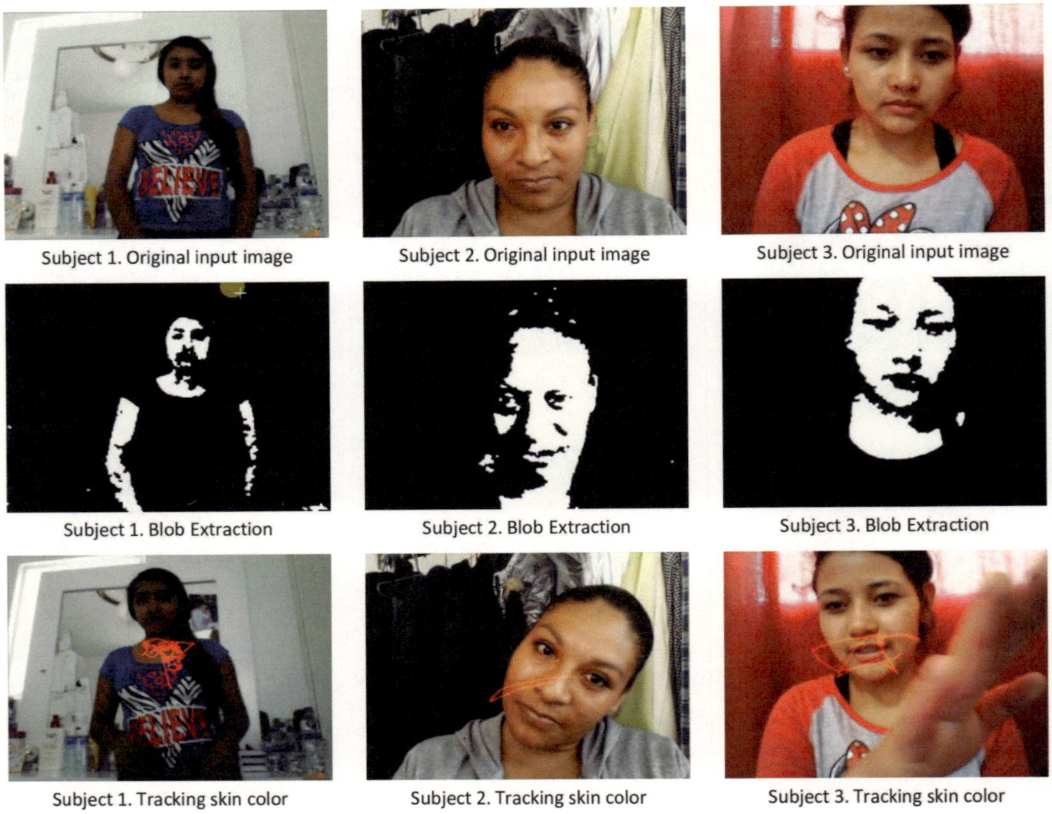

Figure 5. System of detection and tracking of the skin with the technique HS-ab. From top to bottom, the first row displays images temporarily captured by the camera, then in the second, the images can distinguish two types of blobs: black blobs (non-skin regions) and white blobs (skin regions), and the last shows the path (red color) realized by the subject.

4. Conclusion

Numerous application areas use segmentation process such as computer vision and object recognition. Segmentation process is the partitioning of input image or input scene (video scenes) into meaningful objects, and each object can be treated, processed, or discriminated by its salient color. Several color spaces are applied for processing colored images. Each color space has different characteristics and allows to represent the color images. The choice of color space is of the utmost importance and goes according to the application; in this case, our interest is human skin color. There are color spaces for the segmentation of human skin color among the most common HSV, YCbCr, and CIELab [9]. These color spaces consist of chrominance and luminance components, in which the luminance component is affected with lighting changes. This is the reason for the decision to discard the luminance components and use only the chrominance components (H, S, C_b, C_r, a, and b) for skin color segmentation. Based on the above, two techniques are created.

The skin segmentation for color images is a simple and easy way to classify into skin and non-skin pixels. In this work, the skin detection algorithm is used the techniques HS-YCbCr and HS-ab applying histogram threshold-based approach, where the global threshold value for each chrominance component is determined to differentiate pixels from skin color and non-skin color. Later, image preprocessing is required to achieve an efficient segmentation. Morphological operations are the needed operations to complete the segmentation process and reduce noise. There are applied morphological dilation and erosion operations to extract skin pixel. The last stage of the system is the tracking of skin regions. The method of moments is applied to calculate the positions of the largest area of skin color. This method has low computational cost and is easy to implement for the application in computer vision.

This work presents the comparison of two techniques for the modeling of skin color. The experimental results of both skin color segmentation techniques show a good detection of skin color in all images; therefore, the techniques are able to model skin color. However, the technique HS-YCbCr also presents false detections by confusing skin colors with similar ones. With respect to the results of the three metric parameters, the technique HS-ab presents high percentage compared with the other technique indicating that it is a reliable and accurate technique for the detection of skin color pixels. On the other hand, the algorithm for tracking is able to segment skin color in real time.

To end this chapter, a new technique for detection and tracking of skin color has been presented. This proposed system can cope successfully with different colors of skin and complex backgrounds. It can operate with images acquired by a camera and with various body movements.

The future work of research will focus to implement the stage of recognition of gestures in the system and finally test it with amputated people either before or after fixing the prosthesis. Then, this system will be better able to fulfill its main purpose, which consists in developing an effective rehabilitation using the technology of a computer vision system, to return to the patients the highest level of independence and functioning as possible.

Acknowledgements

Authors express their gratitude to Instituto Politécnico Nacional (IPN) and Consejo Nacional de Ciencia y Tecnología de México (CONACYT) for the trust and support to develop research.

Author details

Diana Alejandra Contreras Alejo* and Francisco Javier Gallegos Funes

*Address all correspondence to: dianalecontreras@gmail.com

Instituto Politécnico Nacional (IPN) S.E.P.I Doctorado en Comunicaciones y Electrónica of ESIME Culhuacan, Mexico City, Mexico

References

[1] Bowker J, American Academy of Orthopaedic Surgeons. Atlas of Limb Prosthetics: Surgical, Prosthetic, and Rehabilitation Principles. 2nd ed. Rosemont, IL: Mosby Year Book; 2002. p. 930

[2] Haas D, Somphong P, Jing Y, et al. Kinect based physiotherapy system for home use. Current Directions in Biomecial Engineering. 2015;**1**(1):180-183. DOI: 10.1515/cdbme-2015-0045

[3] Rizzo A, Galen J, Bowerly T, Humphrey L, Neumann U, Rooyen A, Kim L. The virtual classroom: A virtual reality enviroment for the assessment and rehabilitation of attention deficits. Revista Española de Neuropsicología. 2001;**3**(3):11-37

[4] Deyou X. A neural network approach for hand gesture recognition in virtual reality Driving Training System of SPG. In: International Conference on Patter Recognition (ICPR); 20-24 August; Hong Kong, China. IEEE Computer Society; 2006. p. 519-522. DOI: 10.1109/ICPR.2006.109

[5] Yang H, Park A, Lee S. Gesture spotting and recognition for human-robot interaction. IEEE Transaction on Robotics. 2007;**23**(2):256-270

[6] Terrillon J-C, Shirazi M, Fukamachi H, Akamatsu S. Comparative performance of different skin chrominance models and chrominance spaces for the automatic detection of human faces in color images. In: International Conference on Automatic Face and Gesture Recognition; 28-30 March 2000; Grenoble, France. IEEE; 2000. pp. 54-61

[7] Beniak M, Pavlovicova J, Oravec M. Automatic face detection based on chrominance components analysis systems. In: International Conference on Systems, Signals and Image Processing (IWSSIP); 25-28 June 2008; Bratislava, Slovak Republic. IEEE; 2008. pp. 475-478

[8] Chai D, Bouzerdoum A. A bayesian approach to skin color classification in YCbCr color space. In: Region Ten Conference (TENCON); 24-27 September 2000. IEEE; 2000. pp. 421-424

[9] Phung S, Bouzerdoum A, Chai D. Skin segmentation using color pixel classification: Analysis and comparison. IEEE Transactions on Pattern Analysis and Machine Intelligence. 2005;**27**(1):148-154

[10] Aznaveh M, Mirzaei H, Roshan E, Saraee M. A new color based method for skin detection using RGB vector space. In: Conference on Human System Interactions (HSI); May 25-27; Krakow, Poland. IEEE; 2008. pp. 932-935

[11] Vezhnevets V, Sazonov V, Andreeva A. A survey on pixel-based skin color detection techniques. In: International Conference Graphicon; September; Moscow, Russia. 2003. pp. 85-92

[12] Jones M, Rehg J. Statistical color models with application to skin detection. International Journal of Computer Vision. 2002;**46**(1):81-96. DOI: 10.1023/A:1013200319198

[13] Sigal L, Sclaroff S, Athitsos V. Estimation and prediction of evolving color distributions for skin segmentation under varying illumination. In: Conference on Computer Vision and Pattern Recognition (CVPR); 15-15 June 2000; IEEE; 2000. pp. 2152-2159. DOI: 10.1109/CVPR.2000.854764

[14] Fahn CS, Kuo MJ, Wang KY. Real-time face tracking and recognition based on particle filtering and adaBoosting techniques. In: International Conference on Human-Computer Interaction (HCI): Novel Interaction Methods and Techniques. Berlin, Heidelberg: Springer; 2009. p. 198-207. DOI: 10.1007/978-3-642-02577-8_22

[15] Capor A, Tuba M, Vukovik M. Face detection algorithm based on skin detection and invariant moments. In: Recent Advences in Computer Engineering Series; 20-22 February; Cambridge, UK. World Scientific and Engineering Academy and Society Press (WSEAS Press); 2013. p. 110-115. DOI: 10.13140/RG.2.1.3734.8967

[16] Casati J, Moraes D, Rodrigues E. SFA: A human skin image database based on FERET and AR facial images. In: Workshop de Visâo Computacional (WVC); 03-05 June; Rio Janeiro, Brasil. Anais do VIII Workshop de Visão Computacional. 2013

[17] Tabassum M, Gias A, Kamal M, et al. Comparative study of statistical skin detection algorithms for sub-continental human images. Information Technology. 2010;**9**(4):811-817. DOI: 10.3923/itj.2010.811.817

[18] Doukim Ch, Ahmed J, Chekima A. Comparison of three colour spaces in skin detection. Borneo Science. 2009;**24**:75-80

[19] Gasparini F, Schettini R. Skin segmentation using multiple thresholding. Proceedings of SPIE. 2006;**6061**:60610F1-60610F8. DOI: 10.1117/12.647446

For EU product safety concerns:
IN TECH d.o.o., Prolaz Marije Krucifikse Kozulić 3, 51000 Rijeka, Croatia,
info@intechopen.com or visit our website at intechopen.com.